GROUND FAILURES
UNDER SEISMIC CONDITIONS

Proceedings of sessions sponsored by the
Geotechnical Engineering Division of the
American Society of Civil Engineers
in conjunction with the
ASCE National Convention in
Atlanta, Georgia, October 9-13, 1994

Geotechnical Special Publication No. 44

Edited by Shamsher Prakash and Panos Dakoulas

Published by the
American Society of Civil Engineers
345 East 47th Street
New York, New York 10017-2398

ABSTRACT

This proceedings, *Ground Failures Under Seismic Conditions,* contains invited and contributed papers which focus on the behavior of silty and gravelly soils under cyclic loading. Recent case histories, some of which are presented in this volume, reveal that silty and gravelly soils may liquefy and such failures have resulted in landslides, lateral spreading and damages to buildings. The papers present an overview of the dynamic properties, cyclic behavior and liquefaction of silty and gravelly soils. Special emphasis was given in the effect of nonplastic and plastic fine content on the cyclic behavior and strength of saturated silty and gravelly soils. Results have also been presented for partially saturated collapsible soils, demonstrating susceptibility to liquefaction of cyclic softening. Current laboratory and field testing methods have been discussed, including the in-situ soil freezing sampling technique and insitu cyclic testing. Finally, simplified and refined methods of analysis for the prediction of settlements and seismic behavior of soil masses have been examined through a series of case studies.

Library of Congress Cataloging-in-Publication Data

Ground failures under seismic conditions: proceedings of the sessions sponsored by the Geotechnical Engineering Division of the American Society of Civil Engineers in con junction with the ASCE National Convention in Atlanta, Georgia, October 9-13, 1994/edited by Shamsher Prakash and Panos Dakoulas.
    p. cm. — (Geotechnical special publication; no. 44)
    Includes indexes.
    ISBN 0-7844-0055-5
    1. Soil dynamics—Congresses. 2. Seismic waves—Congresses. 3. Soil liquefaction—Congresses. 4. Gravel—Congresses. 5. Silt-Congresses. I. Prakash, Shamsher. II. Dakoulas, Panos. III. American Society of Civil Engineers. Geotechnical Engineering Division. IV. ASCE National Convention (1994: Atlanta, Ga.) V. Series.
TA711.A1G76 1994   94-23246
624.1'51—dc20   CIP

PREFACE

Earthquakes may cause liquefaction of all types of soils below ground level. In the past thirty years, the understanding of liquefaction behavior of sand has dramatically increased. However, recent case histories reveal that silty and gravelly soils also liquefy and these failures have resulted in landslides, lateral spreading and damages to buildings. The behavior of silty and gravelly material is still not well understood. In the past five years many researchers around the world have been involved in research (laboratory, field, and evaluation of failures) to understand the cyclic behavior of silty and gravelly soils.

This Special Technical Publication (STP) constitutes the Proceedings of two sessions held at the ASCE Convention in Atlanta on October 12, 1994 and were sponsored by the Soil Dynamics Committee of the Geotechnical Engineering Division of ASCE. The main objectives of the two sessions were to identify the state of understanding of the liquefaction of silts and gravels, respectively, identify the significant variables, report up-to-date laboratory, field and centrifuge studies and proposed design procedures. World experts from USA, Japan and Canada, presented an overview of the dynamic properties, cyclic behavior and liquefaction of silty and gravelly soils, current laboratory and field testing methods, including in-situ soil sampling techniques, and case histories of liquefaction of silts and gravels.

It is the current practice of the Geotechnical Engineering Division that each paper published in a STP be reviewed for its content and quality. These special technical publications are intended to reinforce the programs presented at convention sessions or specialty conferences and to contain papers that are timely and may be controversial to some extent. Because of the need to have the STP available at the convention, time available for reviews is generally not as long and reviews may not be as comprehensive as those given to papers submitted to the Journal of the Division. Therefore, it should be recognized that there is difference in the purpose and technical status of contributions to the special technical publications as compared to those in the Journal. In accordance with ASCE policy, for these Proceedings, each paper received at least two independent positive peer reviews. All papers published in this volume are eligible for discussion in the Journal of the Geotechnical Engineering Division and are eligible for ASCE awards. Reviews of papers published in this volume were conducted by the Soil Dynamics Committee of the Geotechnical Engineering Division. The following committee members reviewed these papers:

Shahid Ahmad	David Elton	Mladen Vucetic
Shobha K. Bhatia	W.D. Liam Finn	Mishac Yegian
Ahmed W. Elgamal	David Frost	

The session planning was done by myself, Panos Dakoulas, Shobha Bhatia and Mishac Yegian. Thanks are due to them for guidance. I also want to thank the body of experts who gave both the time and effort in reviewing the papers and Shiela Menaker who arranged for the assembly and printing of this volume. Last but not least, thanks are due to all the authors who kindly accepted the invitation and made the most significant contribution to this volume and to the sessions in Atlanta.

Shamsher Prakash, F. ASCE
Professor of Civil Engineering
University of Missouri-Rolla
Rolla, Missouri

Panos Dakoulas, M. ASCE
Associate Professor of Civil Engineering
Rice University
Houston, Texas

Session Organizers and Editors

CONTENTS

GROUND FAILURES UNDER SEISMIC CONDITIONS: SILTY SOILS

GROUND FAILURES UNDER SEISMIC CONDITIONS: GRAVELLY SOILS

GEOTECHNICAL SPECIAL PUBLICATIONS

1) TERZAGHI LECTURES
2) GEOTECHNICAL ASPECTS OF STIFF AND HARD CLAYS
3) LANDSLIDE DAMS: PROCESSES RISK, AND MITIGATION
4) TIEBACKS FOR BULKHEADS
5) SETTLEMENT OF SHALLOW FOUNDATION ON COHESIONLESS
   SOILS: DESIGN AND PERFORMANCE
6) USE OF IN SITU TESTS IN GEOTECHNICAL ENGINEERING
7) TIMBER BULKHEADS
8) FOUNDATIONS FOR TRANSMISSION LINE TOWERS
9) FOUNDATIONS AND EXCAVATIONS IN DECOMPOSED ROCK OF
   THE PIEDMONT PROVINCE
10) ENGINEERING ASPECTS OF SOIL EROSION DISPERSIVE CLAYS
    AND LOESS
11) DYNAMIC RESPONSE OF PILE FOUNDATIONS— EXPERIMENT,
    ANALYSIS AND OBSERVATION
12) SOIL IMPROVEMENT - A TEN YEAR UPDATE
13) GEOTECHNICAL PRACTICE FOR SOLID WASTE DISPOSAL '87
14) GEOTECHNICAL ASPECTS OF KARST TERRIANS
15) MEASURED PERFORMANCE SHALLOW FOUNDATIONS
16) SPECIAL TOPICS IN FOUNDATIONS
17) SOIL PROPERTIES EVALUATION FROM CENTRIFUGAL
    MODELS
18) GEOSYNTHETICS FOR SOIL IMPROVEMENT
19) MINE INDUCED SUBSIDENCE: EFFECTS ON ENGINEERED
    STRUCTURES
20) EARTHQUAKE ENGINEERING & SOIL DYNAMICS (II)
21) HYDRAULIC FILL STRUCTURES
22) FOUNDATION ENGINEERING
23) PREDICTED AND OBSERVED AXIAL BEHAVIOR OF PILES
24) RESILIENT MODULI OF SOILS: LABORATORY CONDITIONS
25) DESIGN AND PERFORMANCE OF EARTH RETAINING STRUCTURES
26) WASTE CONTAINMENT SYSTEMS; CONSTRUCTION, REGULATION, AND
    PERFORMANCE
27) GEOTECHNICAL ENGINEERING CONGRESS
28) DETECTION OF AND CONSTRUCTION AT THE SOIL/ROCK INTERFACE
29) RECENT ADVANCES IN INSTRUMENTATION, DATA ACQUISITION AND
    TESTING IN SOIL DYNAMICS
30) GROUTING, SOIL IMPROVEMENT AND GEOSYNTHETICS
31) STABILITY AND PERFORMANCE OF SLOPES AND EMBANKMENTS II (A
    25-YEAR PERSPECTIVE)
32) EMBANKMENT DAMS-JAMES L. SHERARD CONTRIBUTIONS
33) EXCAVATION AND SUPPORT FOR THE URBAN INFRASTRUCTURE
34) PILES UNDER DYNAMIC LOADS
35) GEOTECHNICAL PRACTICE IN DAM REHABILITATION
36) FLY ASH FOR SOIL IMPROVEMENT
37) ADVANCES IN SITE CHARACTERIZATION: DATA ACQUISITION,
    DATA MANAGEMENT AND DATA INTERPRETATION
38) DESIGN AND PERFORMANCE OF DEEP FOUNDATIONS: PILES AND
    PIERS IN SOIL AND SOFT ROCK
39) UNSATURATED SOILS
40) VERTICAL AND HORIZONTAL DEFORMATIONS OF FOUNDATIONS AND
    EMBANKMENTS
41) PREDICTED AND MEASURED BEHAVIOR OF FIVE SPREAD FOOTINGS
    ON SAND
42) SERVICEABILITY OF EARTH RETAINING STRUCTURES
43) FRACTURE MECHANICS APPLIED TO GEOTECHNICAL ENGINEERING
44) GROUND FAILURES UNDER SEISMIC CONDITIONS
45) IN-SITU DEEP SOIL IMPROVEMENT

LIQUEFACTION OF SILTY SOILS

Y.P. Vaid[1]

ABSTRACT

Basic requirements for a rational assessment of the effect of silt content on the liquefiability of silty sands are outlined. Results from a laboratory study in which these requirements are satisfied are presented in an attempt to ascertain the effect of non plastic silt content on the static and cyclic liquefiability of a given parent sand. The effects of loading direction on static liquefiability of these silty sands are also investigated.

INTRODUCTION

Field performance data during earthquakes show that the soils most susceptible to liquefaction are saturated sands, silty sands and silts. Consequently, liquefaction of these soils has been the topic of extensive laboratory research over the past 25 years. Most of the laboratory studies have, however, been confined to clean sands. The behavior of silty sands and silts has been investigated only on a very limited scale.

---

[1] Professor, Department of Civil Engineering, University of British Columbia, 2324 Main Mall, Vancouver, B.C., Canada V6T 1Z4

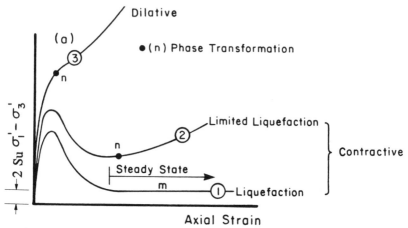

Figure 1. Contractive and dilative responses under static undrained loading.

The term liquefaction and liquefaction failure encompasses all phenomena involving excessive deformations of saturated cohesionless soils (NRC, 1985). Under static loading the term liquefaction is associated with a strain softening type of undrained response, resulting in either unlimited or limited flow deformation (Fig. 1). Sand exhibiting such response is termed contractive (Castro, 1969). If no strain softening occurs, the sand is called dilative. Under cyclic undrained loading, liquefaction can manifest itself either as a strain softening response, much in the same manner as under static loading, or by the development of cyclic mobility (Fig. 2). Cyclic mobility is associated with excursions during loading of the stress state of the sand through transient states of zero effective stress ($\sigma'_3 = 0$), and the sand is deemed to have liquefied if it develops a specified level of strain. The first time occurrence of this $\sigma'_3 = 0$ has been termed initial liquefaction (Seed, 1979). At the conclusion of cyclic loading following liquefaction the residual conditions in the sand are normally assumed to correspond to $\sigma'_3 = 0$, but they need not necessarily be so, depending upon the strain level selected for the definition of liquefaction (Vaid and Thomas, 1994).

This paper described the phenomena of liquefaction in silty sands as assessed in laboratory studies. Previous work is critically reviewed and recent data presented on artificially prepared silty sands, in which particular attention is paid to (i) sample reconstitution by water deposition that simulates natural fluvial and hydraulic fill deposition, (ii) ensure sample homogeneity, an essential requirement in laboratory element tests, (iii) emphasise that moist tamping or air pluviation techniques may yield void ratio states that are not even accessible to the silty sand under water deposition, and (iv) demonstrate that the susceptibility to liquefaction attributed to strain softening phenomenon is dependent on the loading direction.

PREVIOUS WORK

Field Observations

Field observations were instrumental in the first time recognition of the possible effects of silt content on liquefaction resistance. Based on the observed response of sites that liquefied and did not liquefy, Seed et al. (1985) developed correlations between SPT-N value and the cyclic stress ratio to cause liquefaction that clearly recognize the influence of fines content on liquefaction resistance. For a given $(N_1)_{60}$ (the normalized SPT), the liquefaction resistance increased with increasing fines content in excess of 5%. Seed et al. (1988) also proposed similar correlation of residual strength or steady state strength s_u (see Fig. 1) with SPT $(N_1)_{60}$ values based on back-analyses of observed flow failures.

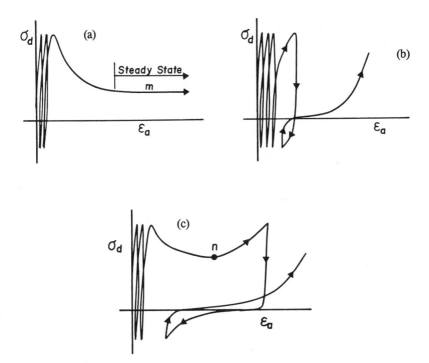

Figure 2.   Strain development under cyclic undrained loading: (a) contractive response of liquefaction type; (b) cyclic mobility; (c) contractive response of limited liquefaction type followed by cyclic mobility.

The field empirical correlation, notwithstanding their great practical importance, however, do not furnish insight into the fundamental mechanical behaviour of liquefaction prone materials. Inherent variability in natural fluvial deposits, hydraulic fills and tailings deposits that constitute the data base for these correlations, provide only average values of liquefaction resistance or residual strength for the entire deposit. A fundamental understanding of their behaviour can only be achieved under controlled laboratory studies as has been customary in any study of material characterisation.

Laboratory Studies

Controlled laboratory studies on liquefaction of silty sands where silt content has been treated as an independent variable for a given parent sand are few. It appears that only Chang et al. (1982) and Troncoso and his associates (1985, 1986) have reported some research in this area. Troncoso and his associates found that at a given void ratio of silty sands the cyclic loading resistance decreased with increase in silt content of up to about 30%. Chang et al. on the other hand, report an increase in cyclic resistance with increase in silt content using the same basis of equal void ratio for comparison.

Test specimens were reconstituted in the laboratory in the aforementioned studies by moist tamping. This technique of sample reconstitution neither simulates the fabric that will ensue on water deposition nor does it guarantee specimen uniformity. Sample homogeneity is tacitly assumed, and normally no direct evidence in support of it is provided. Castro (1969) has reported a variation of as much as 0.04 in void ratio over the height of 7.0 cm moist tamped loose samples of sand. Considering that the residual strength - void ratio relationships for certain sands have a very flat slope, such an error in void ratio can result in several fold variation in residual strength, and hence the results of such tests of questionable validity. The moist tamping technique, furthermore, enables sample reconstitution at void ratios that may not even be accessible to the soil in the water deposited state.

ESSENTIAL REQUIREMENT IN LABORATORY STUDIES

Fundamental laboratory studies of material behaviour require tests on homogeneous specimens under uniform state of stress. In addition, the specimen reconstitution method should simulate the mode of deposition of the soil deposit being modelled. Water pluviation technique for sample reconstitution has been considered to simulate fluvial and hydraulic fill sand fabrics (Oda et al., 1978). This technique, however, results in segregation when used with silty or well-graded sands. Kuerbis and Vaid (1988) have developed a slurry method of deposition that yields homogeneous specimens of well-graded as well as silty sands. This technique of placement simulates closely the deposition through water of hydraulic fills and fluvial sands, and thus provides a convenient means for a systematic study of their undrained behaviour. Homogeneity of laboratory test specimens is mandatory in order to determine element properties. It is not implied herein that hydraulic fills and fluvial sands may be homogeneous over large thicknesses. Since laboratory tests must be element tests, the technique used by Vasquez et al. (1988) of water pluviation of silty sand, with time interruptions that gave rise to a 5 layered material are fundamentally incorrect. The results from such types of tests will depend upon the specimen size used.

HOMOGENEITY OF SLURRY DEPOSITED SILTY SANDS

The homogeneity of the slurry deposited triaxial samples (63 mm diam. x 126 mm high) is assessed by comparing grain size distributions of several horizontal slices. Horizontal slices were cut after the specimen had been solidified using the gelatine technique (Emery et al. 1973). Figure 3 shows that

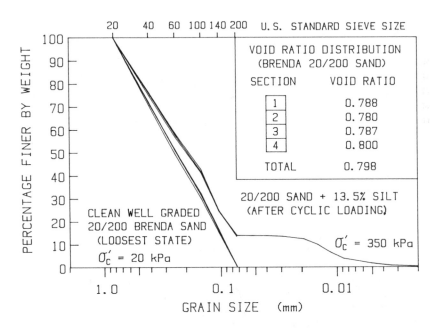

Figure 3. Homogeneity of slurry deposited sand and silty sand samples.

there is virtually no difference among the grain size distribution of slices either for well-graded Brenda 20/200 clean or silty sand specimens, implying a high degree of specimen homogeneity. A similar test on a water pluviated specimen of 20/200 sand clearly shows considerable segregation (Fig. 4). Brenda sand is an angular tailing sand. The designation 20/200 implies processed sand that passes #20 and is fully retained on #200 sieves, and possesses a straight line gradation.

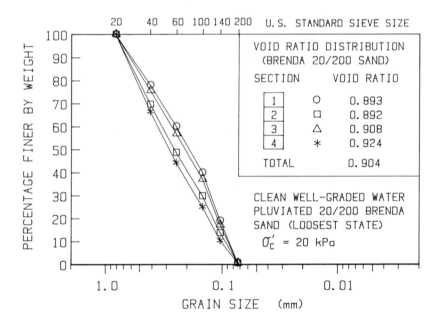

Figure 4.   Homogeneity of water pluviated sand samples.

ACCESSIBLE AND INACCESSIBLE STATES

Loosest (maximum) void ratios accessible to Brenda 20/200 silty sands are illustrated in Fig. 5. The maximum void ratio of a silty sand deposited by slurry deposition depends on silt content. It may be seen to decrease approximately linearly as silt content increases form zero to about 20%. In contrast, maximum void ratio determined by the ASTM method is approximately equal to that of the

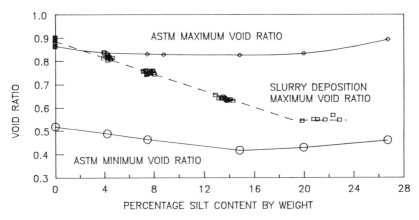

Figure 5.   ASTM dry maximum and minimum void ratios and slurry deposition
             maximum void ratios of silty 20/200 Brenda sands.

clean sand and is essentially independent of silt content up to about 20%.
Although ASTM method for maximum and minimum void ratios is not
recommended for silt content in excess of 12%, the data in Fig. 5 extends to silt
contents up to 27%. This was intended for direct comparison of void ratios
obtained by air pluviation and the new slurry deposition techniques over a large
range of silt contents. It is clear from Fig. 5 that silty sands get placed in much
denser states by slurry deposition than by air pluviation. Loose densities on air
pluviation, in fact, are not possible for silty sand on slurry deposition. Similarly
silty sands can be reconstituted at any arbitrary void ratio by moist tamping,
despite the fact that such a void may not even be accessible under water
deposition.   Both air pluviation and moist tamping technique of reconstitution
neither simulate the deposition process nor the range of void ratios possible in
fluvial or hydraulic fill sands.

STATIC LIQUEFACTION OF BRENDA SILTY SANDS

        Undrained behaviour of several silty sands comprised of parent Brenda
20/200 sand with different silt (nonplastic) contents is shown in Fig. 6(a). Test
samples were hydrostatically consolidated under σ'_3 = 350 kPa following loosest
deposition by the slurry method.   Figure 6(a) shows that the behaviour in
compression is dilative and there is a trend towards more dilative behaviour with
increasing silt content. In contrast, the behaviour in extension is contractive and
shows only a slight tendency towards reduced contractiveness with increasing silt
content.   Apparently silt content affects response in compression loading .
Nevertheless, the silty sands are anisotropic in their undrained response -
liquefiable in extension but not in compression even in the loosest deposition

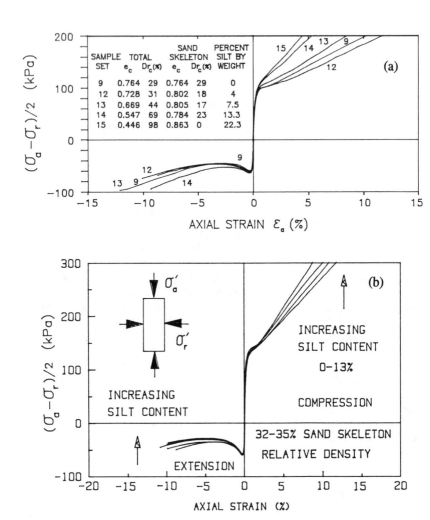

Figure 6.   Static strain softening and dilative undrained response of (a) Brenda
20/200 and (b) Brenda 20/40 silty sands.

state. This inherent anisotropy in undrained response is characteristic to pluviated sand that has been demonstrated by several researchers (e.g. Bishop, 1971; Chang et al., 1982; Hanzawa, 1980; Vaid et al., 1989). Experimental evidence that demonstrates a systematic weakening of undrained response (changing from dilative to contractive) of sands as the loading direction (σ_1 direction), changes from vertical (compression) to horizontal (extension) have been given by Symes et al. (1985) and Shibuya and Hight (1987). Consequently, Seed et al (1988) SPT-residual strength correlations must represent residual strength values that are averaged for several loading directions inherent in a sloping ground failure.

Silty sands comprised of the uniform coarser Brenda sand 20/40 may be seen to have behaviour similar to that of well-graded Brenda 20/200 silty sands (Fig. 6b), implying no fundamental difference in behaviour with change in gradation or average grain size.

Since the void ratio following slurry deposition decreases substantially with silt content, the response shown in Fig. 6(a) represents very diverse void ratios. It has been suggested that relative density is not a suitable index for characterizing behaviour of silty sands (Ishihara et al., 1980). The results in Fig. 6(a) wherein silty sand (22.5% silt content) at 98% relative density (based on ASTM maximum and minimum void ratios) is seen to be contractive, and has only a slight difference in behaviour from that of clean sand at 25% relative density, gives further support to this contention. Void ratio instead of relative density has been used for comparing the behaviour of silty materials by Ishihara et al. (1980). The substantial decrease in void ratio with increasing silt content for samples which have only a minor change in behaviour suggests that the use of void ratio may also be inadequate for characterizing the behaviour of silty sands. It is possible that the sand skeleton void ratio e_s instead of the total void ratio may govern the behaviour of silty sands. The sand skeleton void ratio (Kenny, 1977) is defined by

$$e_s = V_T G_s \rho_\omega / (M - M_{silt}) - 1 \qquad (1)$$

in which V_T = total volume of the specimen, G_s = specific gravity of solids, ρ_ω = density of water, M = total mass of solids and M_{silt} = mass of silt. The concept of e_s implies that the silt simply occupies the interstices formed by the sand grains and thus the silty sand response is governed essentially by the sand skeleton void ratio.

Evidence in support of this postulate may be seen in Fig. 7 which is plotted using data in Fig. 5. It may be seen that on slurry deposition various silt

contents give rise to an essentially constant sand skeleton void ratio, even though
the actual void ratio of the total soil decreases markedly with increase in silt
content. Air deposition, on the other hand, may be seen to increase greatly the
sand skeleton void ratio. This is a clear illustration of the bulking effect of silt on
the void ratio of sand in the dry state.

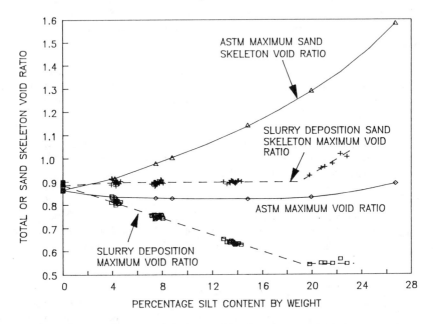

Figure 7.   Maximum sand skeleton void ratios of Brenda 20/200 silty sand on
air pluviation and slurry deposition.

        When silty sand is water deposited from a slurry state, the sand fraction
tends to settle more quickly than the silt fraction due to its higher terminal
velocity (Vaid and Negussey, 1988). The deposition process should be similar to
pluviation of clean sand through water, unless the silty slurry is excessively thick,
which will occur at high silt contents. As a result, the maximum sand skeleton
void ratio on deposition through silty slurry are likely to vary little with silt
content, as shown by test data in Fig. 13 for silt content up to about 20%. For
larger silt contents, the silt slurry through which sand pluviates is very thick and
viscous. Consequently, e_s increases.

LIQUEFACTION UNDER CYCLIC LOADING

Figure 8 illustrates liquefaction resistance curves at σ'_{3c} = 350 kPa for the slurry deposited Brenda 20/200 silty sands. These resistance curves show cyclic stress ratio $\sigma_d/2\sigma'_{3c}$ to cause liquefaction (defined herein as the development of 2.5% single amplitude axial strain in 10 stress cycles). The curves reflect only the cyclic mobility form of liquefaction failure, because the cyclic stress amplitude never exceeded the steady state or phase transformation strength in either compression or extension mode (Vaid et al., 1989).

It may be noted in Fig. 8 that silty sands at the loosest state have similar cyclic strength regardless of silt content, even though there is a large variation in void ratios. The resistance curves are seen to essentially shift horizontally towards lower void ratios as silt content increases. At each silt content, there appears to be an approximately similar increase in cyclic resistance for a given decrease in void ratio. Over the range of overlapping void ratios of the various silty sands, the cyclic resistance at a given void ratio decreases with increase in silt content. Such a comparison is not possible at all void ratios, since certain void ratios are not accessible to some water deposited silty sands.

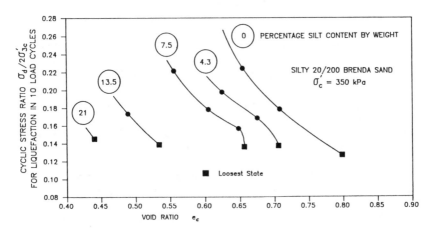

Figure 8.   Resistance to liquefaction of silty 20/200 sand as a function of consolidation void ratio.

An alternative comparison of cyclic resistance of silty sands as a function of ASTM relative density is shown in Fig. 9. The ASTM relative density curves show a similar relationship between density and cyclic resistance as void ratio

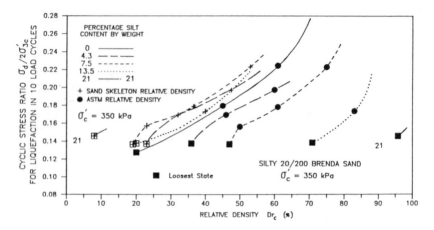

Figure 9.   Resistance to liquefaction of silty 20/200 sand as a function of
relative density.

curves of Fig. 8, i.e. a higher silt content results in a lower cyclic resistance for a
given relative density or void ratio.

Cyclic loading resistance in terms of sand skeleton relative density is also
shown in Fig. 9.   Sand skeleton relative density was calculated using e_s together
with maximum and minimum ASTM void ratios for the sand fraction.   Although
in terms of ASTM relative density, a higher silt content results in a lower cyclic
resistance for a given relative density or void ratio, comparison in terms of sand
skeleton relative density brings all resistance curves very close to each other.
These curve show a slight increase in cyclic resistance with increase in silt
content at a given value of sand skeleton void ratio.   It would thus appear that the
cyclic strength of a silty hydraulic fill sand could be determined by running tests
on clean sand at identical sand skeleton void ratio of the silty sand   being
modelled.   This strength estimate obtained would tend to be conservative.

Figure 10 illustrates a range of cyclic loading behaviour of silty sands at
various fixed sand skeleton relative densities.   These curves are based on results
from over 80 cyclic tests.   It may be noted that at each sand skeleton density,
increase in silt content leads to some increase in cyclic stress ratio to cause
liquefaction in a fixed number of cycles.

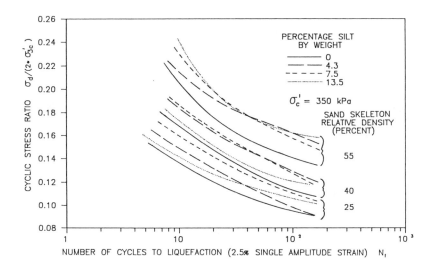

Figure 10. Summary of the variation of cyclic strength of silty 20/200 sand with silt content at constant sand skeleton relative density.

CONCLUSIONS

A rational assessment of the effect of silt content on the liquefiability of silty sands can only be made in controlled laboratory studies in which
(i)   silt content is varied in a given parent sand
(ii)   uniformity of test specimens is ensured which is mandatory in all element tests and
(iii)   specimens are reconstitued by a deposition process that simulates the water deposition processes of fluvial and hydraulic fill silty sands that have been found prone to liquefaction.

Test results from such a laboratory study in which a broadly graded (Cu≈6) parent sand was mixed with an organic silt varying from 0 - 22% show that
• Water deposited silty sands at a given density and confining stress may be dilative in triaxial compression yet contractive in extension. Steady state or residual strength for a given silty sand is thus not unique but depends on the loading direction.
• Increase in silt content up to 20% in the sand tested makes compression behaviour somewhat more dilative and extension behaviour less contractive.

The rather minor change in behaviour with increasing silt can be explained by the fact that sand skeleton void ratio remains virtually unaltered with the addition of silt. Silt could be considered as an essentially inert component which fills sand skeleton voids.

* Silty sands in the loosest slurry deposited state possess similar cyclic strength despite wide variation in void ratio. However, when compared at constant sand skeleton void ratio, cyclic strength improves slightly with increase in silt content.

ACKNOWLEDGEMENTS

This research was supported by a grant from the Natural Science and Engineering Research Council of Canada. The data on Brenda silty sands was obtained by Ralph Kuerbis during his MASc research under the author's supervision. Kelly Lamb prepared the manuscript.

REFERENCES

1. Bishop, A.W. (1971). Shear strength parameters for undisturbed and remoulded soil specimens. Roscoe Mem. Symp., Cambridge University, pp. 3-58.

2. Chang, N.Y., Yey, S.T. and Kaufman, L.P. (1982). Liquefaction potential of clean and silty sands. Proc. 3rd Microzonation Conf., Seattle, pp. 1018-1032.

3. Castro, G. (1969). Liquefaction of sands. Ph.D. Thesis, Harvard Univ., Cambridge, Mass.

4. Emery, J.J., Finn, W.D.L. and Lee, K.W. (1973). Uniformity of saturated sand samples. ASTM STP 523, pp. 182-194.

5. Hanzawa, H. (1980). Undrained strength and stability of quick sand. Soils and Foundations, Vol. 20, (2), pp. 17-29.

6. Kenny, T.C. (1977). Residual strength of mineral mixtures. Proc. 9 ICSMFE, Tokyo, Vol. 1, pp. 155-160.

7. Ishihara, K., Troncoso, J., Kawase, Y. and Takahashi, Y. (1980). Cyclic strength characteristics of tailings materials. Soils and Foundations, Vol. 20 (4), pp. 127-142.

8.   Kuerbis, R.H. and Vaid, Y.P. (1988). Sand sample preparation - the slurry deposition method. Soils and Foundations, Vol. 28 (4), pp. 107-118.

9.   National Research Council (1985). Liquefaction of soils during earthquakes. Report No. CETS-EE-001, Com. Earthq. Eng., National Academy Press, Washington, D.C.

10. Oda, M., Koishikawa, I. and Higuchi, T. (1978). Experimental study of anisotropic shear strength of sand by plane strain test. Soils and Foundations, Vol. 18 (1), pp. 25-38.

11. Seed, H.B. (1979). Soil liquefaction and cyclic mobility evaluation for level ground during earthquakes. ASCE Journal of Geotech. Eng., Vol. 105, GT2, pp. 201-255.

12. Seed, H.B., Tokimatsu, K., Harder, L. and Chung, R.M. (1985). Influence of SPT procedures in soil liquefaction resistance evaluation. JGE, ASCE, Vol. 103.

13. Seed, H.B., Seed, R.B., Harder, L. and Jong, H. (1988). Re-evaluation of the slide in the lower San Fernando Dam in the earthquake of February 9, 1971. Report UCB/EERC-88/04.

14. Shibuya, S. and Hight, D.W. (1987). A bounding surface for granular materials. Soils and Foundations, Vol. 27 (4), pp. 123-136.

15. Symes, M.J. Shibuya, S., Hight, D.W., and Gens, A. (1985). Liquefaction with cyclic principal stress rotation. 11th International Conference on Soil Mechanics and Foundation Engineering, Vol. 4, San Francisco.

16. Troncoso, J.H. (1986). Critical state of tailings silty sands for earthquake loading. Soil Dynamics and Earthquake Engineering, Vol. 5, pp. 248-252.

17. Troncoso, J.H. and Verdugo (1985). Silt content and dynamic behaviour of tailings sands. Proc. 11, ICSMFE, San Francisco, Vol. 3, pp. 131--1314.

18. Vaid, Y.P. and Negussey, D. (1988). Preparation of reconstituted sand specimens. ASTM STP 997, pp. 405-420.

19.   Vaid, Y.P., Chung, E.F.K. and Kuerbis, R. (1989). Preshearing and undrained response of sand. Soils and Foundations, Vol. 29, pp. 49-61.

20.   Vaid, Y.P. and Thomas, J. (1994). Post-liquefaction behaviour of sands. Proc., 13th ICSMFE, New Delhi, Vol. 1.

21.   Vasquez-Herrara, A. and Dobry, R. (1988). The behaviour of undrained contractive sand and its effect on seismic liquefaciton flow failure of earth structures. RPI, Troy, N.Y.

THE INFLUENCE OF FINES TYPE AND CONTENT
ON CYCLIC STRENGTH

by Joseph Patrick Koester[1], Member ASCE

Abstract

Many critical facilities, such as large dams, bridges, and power plants, are founded on either alluvial or artificially reclaimed saturated deposits of low- to medium plasticity sandy silts, clayey silts, or other mixture soils that may be susceptible to drastic loss of shear strength or liquefaction consequent to earthquake shaking. Soils containing plastic fines were considered resistant to development of high residual excess pore water pressures leading to strength loss until such soils were observed to have liquefied during the 1976 Tangshan, People's Republic of China, earthquake. Recent reexamination of fine-grained soil behavior during earthquakes and results of laboratory tests reveal that uniformly graded loose sandy soils that contain as much as 25-30% fines may be highly liquefiable.

A series of nearly 500 undrained cyclic triaxial tests was conducted on reconstituted mixtures of sand, silt, and plastic clay to define the influence the gradation and index properties of the fines fraction on liquefaction resistance and pore pressure generation characteristics. A companion series of undrained hollow cylinder cyclic torsional shear tests was also performed on mixtures selected from the cyclic triaxial test program to investigate their behavior when subjected to cyclic simple shear and very large monotonic shear strains. The results of the laboratory program are presented and indicate that the lowest liquefaction resistance in mixtures prepared to a unique global void ratio (corresponding to 50% relative density of the sand fraction before addition of

[1]Research Civil Engineer, US Army Engineer Waterways Experiment Station, 3909 Halls Ferry Road, ATTN: CEWES-GG-H, Vicksburg, Mississippi 39180-6199

fines) occurs at fines contents between 20% and 26%. Plasticity of the fines exerts a less pronounced strengthening effect.

Monotonic, undrained torsional shear tests were performed both on hollow cylindrical specimens of silty soils that had been liquefied by cyclic simple shear, and replicate specimens that had not been subjected to any previous loading. Residual strengths of specimens containing 20% low-plasticity fines were found to be very low; deformation potential in such soils would be essentially unlimited.

Background

Many critical facilities, such as large dams, bridges, and power plants, are founded on either alluvial or artificially reclaimed saturated deposits of low- to medium plasticity sandy silts, clayey silts, or other mixture soils that may be susceptible to drastic loss of shear strength or liquefaction consequent to earthquake shaking. Soils containing plastic fines were considered resistant to development of high residual excess pore water pressures leading to strength loss until such soils were observed to have liquefied during the 1976 Tangshan, People's Republic of China, earthquake (Wang 1979, 1981). Recent reexamination of fine-grained soil behavior during earthquakes and results of laboratory tests reveal that uniformly graded loose sandy soils that contain as much as 25-30% fines may be highly liquefiable (Chang 1990).

Laboratory cyclic testing program

General. A two-part laboratory testing program was conducted on saturated artificial mixtures of sand, silt, and clay to evaluate the effects of gradation and index properties on cyclic strength and pore pressure generation characteristics. Fine-grained soils were considered for the purposes of the study as soil materials that contain appreciable fines, yet do not classify as a clay (CL or CH) in the Unified Soil Classification System. Soil fines were defined as those soil materials passing a US Standard No. 200 sieve (silts, or clays, or mixtures of both, finer than 0.074 mm). Two fundamentally different experimental procedures were employed to evaluate undrained cyclic loading resistance of the various mixtures studied: (1) cyclic triaxial (conventional, i.e., constant σ_3, sinusoidally varying σ_1 about $\sigma_1 = \sigma_3$) tests on solid cylindrical specimens and (2) cyclic torsional shear tests on hollow cylindrical specimens. Both types of cyclic loading were stress-controlled.

Table 1 is a matrix of soil mixtures prepared from a uniform fine sand and preselected proportions of a uniform, low-plasticity silt (Vicksburg, Mississippi loess) and plastic clay (Vicksburg, Mississippi "buckshot").

Table 1. Cyclic triaxial test matrix[1] for specimens with fine parent sand (void ratio of parent sand at D_r = 50%, e_{50} = 0.728)

P. I. of Fines (%)[2]	Fines Content of Specimen, %					
	5	12.5	20	30	45	60
4	F22	F32	F42	F52[3]	F62	F72
10	F23	F33	F43	F53	F63	F73
15	F24	F34	F44	F54	F64	F74
20	F25	F35	F45	F55	F65	F75
25	F26	F36	F46	F56	F66	F76
30	F27	F37	F47	F57	F67	F77
40	F28	F38	F48	F58	F68	F78

[1] a. The format for abbreviated codes in this and similar matrixes for medium and well-graded sand mixtures is LN_1N_2, where: L is M, F, or W for medium, fine, or well-graded parent sand, respectively; N_1 ranges from 2 through 7, respectively representing fines contents of 5%, 12.5%, 20%, 30%, 45%, and 60%; and N_2 ranges from 2 through 8, respectively representing plasticity indexes of 4%, 10%, 15%, 20%, 25%, 30%, and 40%.
    b. The three parent sands, i.e., M11, F11, and W11 were tested, as well.
    c. Each tabulated soil type was tested using effective consolidation pressures of 15 and 30 psi (103.4 and 206.8 kPa, respectively).
[2] See Figure 2 for % Buckshot clay (CH).
[3] Underlined soil types were not tested based on trends in previous test results that indicated these could be eliminated without reducing data base coverage.

Sufficient weights of dry constituents for all tests of a given mixture were blended with water to achieve a moisture content of about 7.5%. Moistened soil was stored in plastic bags in a humid room until needed for specimen preparation. Specimens were prepared by moist compaction (tamping of equal-weight lifts to achieve desired lift thicknesses; five for cyclic triaxial tests, 10 for hollow cylinder tests) for both parts of the testing program from mixtures represented in this matrix and similar matrixes for well-graded and medium, uniform parent sands. All specimens were compacted and backpressure saturated to achieve a post-consolidation global void ratio corresponding to 50% relative density when the specimen was constructed only of one of the three "parent" sand gradations (termed e_{50}). Parent sand gradations are charted in Figure 1, along with that of a concrete sand that

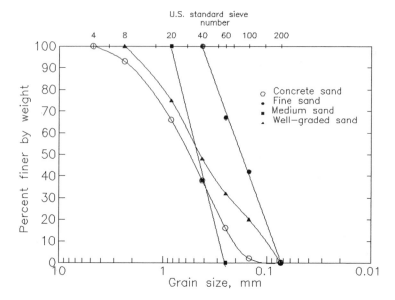

Figure 1. Gradations of supplied concrete sand and test parent sands

served as a supply for most of the sand fractions required to generate the fine, medium, and well graded sands.

Plasticity of the fines fraction (% passing #200 sieve) of the soil mixtures tested was controlled by varying the proportion of clay ("Buckshot") to silt in each mixture. Figure 2 shows the relationship developed for various clay/silt ratios for this program. Plasticity of the fines fraction was found to be a useful control variable for laboratory testing; small quantities of sand cause large variation in Atterberg limits, and the -#200 fraction is directly available from sieve analysis.

Cyclic triaxial tests. A series of nearly 500 undrained cyclic triaxial tests was conducted on isotropically-consolidated specimens prepared from a possible 129 reconstituted mixtures of sand, silt, and plastic clay to develop a data base on the influence of gradation and index properties of the fines fraction on liquefaction resistance and pore pressure generation characteristics. Three principal plotting formats were used to present the large quantity of test results obtained in this test program: cyclic triaxial strength (expressed as the cyclic triaxial stress ratio causing either 100% residual excess pore pressure response or 2.5% double amplitude cyclic axial strain, whichever occurred

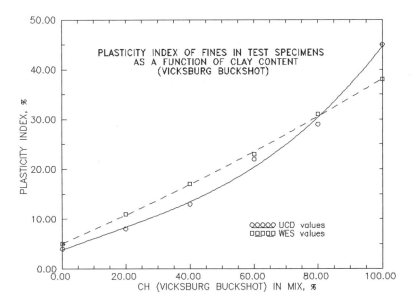

Figure 2. Fines plasticity index variation with clay content (Chang, 1990)

first) as a function of the number of applied load cycles; cyclic stress ratio
associated with 10, 30, and 100 cycles to initial liquefaction (i.e., 100%
residual excess pore pressure response) as a function of fines content; and
cyclic stress ratio associated with 10, 30, and 100 cycles to initial liquefaction
(i.e., 100% residual excess pore pressure response) as a function of plasticity
index of the fines fraction of each mixture as justified above. Although it is
not appropriate or practical to present all such plots generated by the test series
in a conference proceedings paper, several examples will illustrate the trends
observed. Comprehensive results are given by Koester (1992) and are
available from University Microfilms International, Ann Arbor, Michigan.

Figure 3 presents results according to the first plotting format described
above and the series symbol (the five-letter title of the plot) code described as
follows:

(a) The first character delimits the parent sand; M for medium, as
depicted, W for well-graded, and F for fine.
(b) The second character defines the fines content: a numeral identifies
the fines content of all specimens represented in the plot, following the
convention defined in Table 1; the letter X signifies that each curve

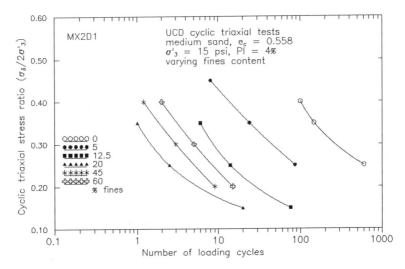

Figure 3. Cyclic triaxial strength curves (stress ratio versus number of cycles to initial liquefaction) for medium sand mixtures, with index properties and test conditions shown (after Chang, 1990)

relates to a particular fines content, according to the legend at the upper right, which in turn follows the convention given in Table 1.

(c) The third character indicates whether plasticity index of the fines fraction is constant (if a numeral, following the convention of Table 1) or if each of curve relates to a particular plasticity index (i.e., if an X) in accordance to the legend at upper right.

(d) The fourth character is either D or L, depending, respectively, on whether the post-consolidation void ratio of all specimens represented is the same as that of its parent sand at 50% relative density or 30% relative density (the latter was examined in a few cases not discussed in this paper).

(e) The fifth character is either 1 or 3, depending, respectively, on whether the effective confining stress to which all specimens were consolidated was 15 psi (103.4 kPa) or 30 psi (206.8 kPa).

For example, Figure 3 represents a comparison of test results obtained on specimens prepared from medium sand (M), varying fines content (X) for which the fines fraction has a constant plasticity index of 4% (corresponding to the numeral 2, see Table 1), specimen void ratios correspond to that of the parent sand at 50% relative density (D), and consolidated under an effective confining stress of 15 psi, or 103.4 kPa (1).

The latter code format, used for plots where failure was associated with a given number of load cycles, consists of the same five characters described above, with the additional numerals 10, 30, or 100 to signify occurrence of initial liquefaction in the respective number of load cycles.

In the case of Figure 3, no "Buckshot" clay was present in the mixtures represented, thus the plasticity index of the soil fines (4%) is that of Vicksburg loess (silt) comprising the fines. The highest cyclic strengths were exhibited by the clean medium parent sand itself. Cyclic strength decreased with addition of low plasticity silt fines, until a fines content of 20% (by weight) was reached; cyclic strengths increased as silt in the mixtures exceeded 20%. Throughout the entire cyclic triaxial test program, in fact, it was observed that a particular fines content corresponded to a lower bound cyclic strength. Figures 4 and 5 show similar results with mixtures based on fine and well-graded parent sands, respectively, with the exception that the first 5% of added fines was observed to accompany an increase in cyclic strength. The lowest fines content associated with lower bound cyclic triaxial strengths were generally observed in specimens constructed from well-graded parent sands. Consistently, the lowest cyclic triaxial strength was measured in gradations containing between 20% and 30% (by weight) silt or silt and clay fines.

Membrane compliance, manifested by the penetration of the confining membrane into peripheral voids of uniformly graded sands when undrained confining pressure is applied, may lead to overestimation of undrained cyclic strength. The mean grain size of the medium parent sand used in these studies was about 0.48 mm. Sands this coarse have been shown to be subject to overestimation of cyclic triaxial strength against liquefaction in 30 loading cycles by as much as 35% (Martin, Finn, and Seed 1978). The conspicuously high cyclic triaxial strengths observed in clean medium sands may have resulted, in part, due to such compliance. As fines were added and filled the voids between the parent sand particles, however, the potential for membrane penetration compliance was reduced. By similar logic, it was considered unlikely that cyclic triaxial strengths of specimens constructed using fine or well-graded parent sands in this study were affected by membrane penetration compliance.

Figure 6 depicts the resistance of specimens constructed from fine parent sand to 30 cycles of either axial or torsional load at various stress ratios, and illustrates the influence of fines content and plasticity of fines on this strength. Plasticity of fines exerts an inconsistent effect on cyclic

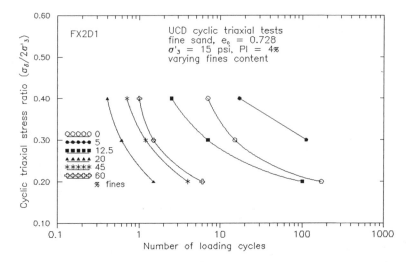

Figure 4.   Cyclic triaxial strength curves (stress ratio versus number of cycles to initial liquefaction) for fine sand mixtures, with index properties and test conditions shown (after Chang, 1990)

strength; fines content affects most gradations in a similar manner to that described for the plot in Figure 3. Fines contents accompanying lower bound cyclic triaxial strengths were generally found to be slightly higher when strengths were interpreted at higher numbers of load cycles.

Hollow cylinder torsional shear tests. A companion series of undrained cyclic torsional shear tests was also performed on mixtures selected from the cyclic triaxial test program to investigate their behavior when subjected to cyclic simple shear and very large monotonic shear strains. A hollow cylinder testing apparatus, designated HCTA-88, developed at the University of Colorado at Denver (UCD) Geotechnical Laboratory and used in this program, is described by Chen (1988). The HCTA-88 test apparatus, equipped with a high-capacity Instron™ cyclic axial/torsional loader, was used to apply cyclic and monotonic undrained simple shear to hollow cylindrical specimens of four moist-compacted sand, silt, and clay mixtures. Torsional cyclic and monotonic loading and data acquisition systems developed for and employed in this study using the UCD apparatus are detailed by Koester (1992). The four soil types were selected from the fine sand cyclic triaxial test matrix (Table 1): clean fine sand (F11); fine sand mixed with 20% fines having a plasticity index of 10 (F43); fine sand mixed with 20% fines having a plasticity index of

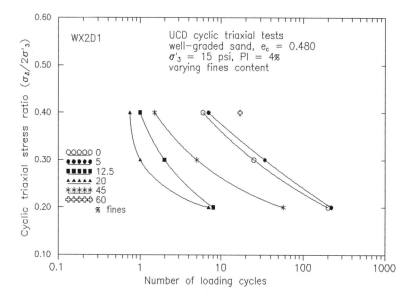

Figure 5. Cyclic triaxial strengths (stress ratio versus number of cycles
to initial liquefaction) for well-graded sand mixtures, with
index properties and test conditions shown (after Chang, 1990)

25 (F46); and fine sand mixed with 45% fines having a plasticity index of 15
(F64).

Applied torsional simple shear stress is plotted against effective
confining stress in Figure 7 as measured during a test on clean fine sand to
illustrate typical features observed in most of the specimens under undrained
cyclic torsional simple shear loading. The load- (or stress-) control nature of
the test is evident from the constant peak applied shear stress attained in each
load cycle, until the failure envelope was reached. Excess pore pressures
developed in this specimen with the first load cycle and continued, at a more
or less steady rate, until the failure envelope was approached. Once about
70% of the effective confining stress was lost due to this pore pressure
buildup, leftward migration of the applied cyclic shear stress path so plotted
accelerated to failure within a few cycles.

The failure envelopes drawn to fit the applied cyclic shear stress path in
Figure 7 at the final few cycles of loading are symmetric to the horizontal
(zero shear stress) axis. Applied shear stress exerts no bias on the response of

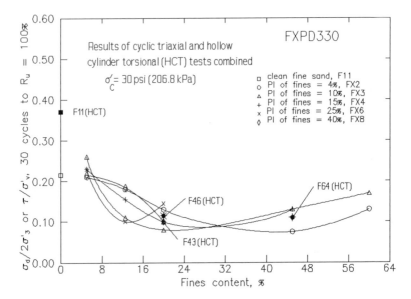

Figure 6. Cyclic stress ratios causing initial liquefaction in 30 cycles in fine sand mixtures. Void ratio equals that of parent sand at D_r = 50% (after Chang, 1990)

isotropically consolidated specimens when there is no initial static shear stress. This is not the case with stress-reversal cyclic triaxial tests, where the failure envelope associated with extension of the specimen is always, at least temporarily, reached before the compression failure envelope.

A "triggering" pore pressure ratio of about 70% leading to liquefaction has been observed in centrifuge experiments (Steedman and Schofield, 1991). Slightly higher excess pore pressures were found to be required to trigger a liquefaction response in soils containing more than 20% fines in cyclic torsional simple shear tests in this study. Zhu and Law (1988) observed that 80% excess pore pressure ratios triggered liquefaction response in reconstituted specimens of nonplastic silt in undrained cyclic triaxial tests, and that undisturbed specimens of the same silt liquefied whenever 85% pore pressure ratio was attained. Acceleration of cyclic shear strain increase is more abrupt in soils with higher fines; the F43, F46, and F64 soils failed in a more "brittle" fashion than did the clean sand, beginning at higher excess pore pressure ratios (about 80%).

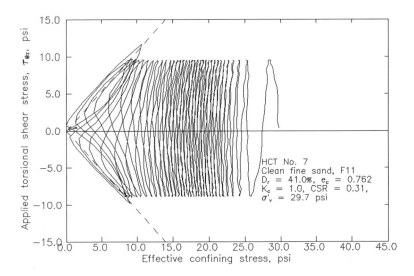

Figure 7. Effective stress path based on applied torsional shear stress, HCT Test No. 7

Cyclic strength curves developed for hollow cylinder torsional simple shear tests on all four mixtures conducted at 30 psi (206.8 kPa) effective confining stress are combined in Figure 8. The trends observed in the companion cyclic triaxial test program hold, namely: the addition of up to 20% fines reduces cyclic strength behavior to a lower bound value in these mixtures at a single post-consolidation void ratio; cyclic strength increases beyond the lower bound with additional soil fines; and plasticity index of the fines fraction exerts a lesser influence on cyclic strength behavior than does gradation. Strengths defined for 30 cycles of loading are plotted in Figure 6 for comparison with cyclic triaxial test results.

Laboratory residual strength testing

General. Monotonic, undrained torsional simple shear tests were performed both on hollow cylindrical specimens of silty soils that had been liquefied by cyclic simple shear, and replicate hollow cylinder specimens that had not been subjected to any previous loading. The purpose for these tests was to measure the post-liquefaction residual strength of fine-grained mixture soils and evaluate the effects, if any, of fines content and index properties on

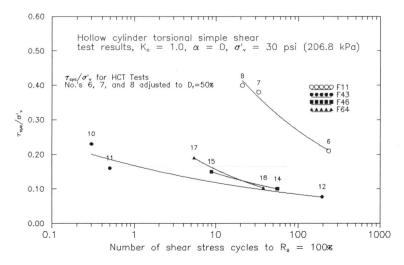

Figure 8.   Cyclic strengths of HCT specimens isotropically consolidated
to 30 psi (206.8 kPa) effective confining stress

residual strength. To test previously liquefied specimens, the specimens were returned to a zero-strain condition (i.e., the top of the specimen was gradually returned to its pre-cyclic test orientation) with drainage valves left closed and a linear ramp function of torsion was applied until the limits of the apparatus was reached, corresponding typically to about 25% shear strain. Replicate virgin specimens of each of the four fine sand-based mixtures were also tested in monotonic torsional shear; with the top platen of the HCTA-88 apparatus properly placed during specimen preparation, it was possible in these tests to achieve about 40% shear strain before the limits of the device were reached. Shear strain was applied at a rate of about 1.15% per minute in these monotonic tests.

Membrane torque correction. Shearing stresses are developed in the confining membranes lining the inner and outer cylindrical surfaces of hollow cylinder specimens as they are subjected to rotational deformation (twist). Given the dimensions of specimens tested in this research, and the use of specially manufactured latex membranes that were thicker than typical triaxial test membranes to provide durability for compacted specimen preparation, it was decided to investigate the magnitude of stresses resulting from membrane stiffness. The primary objective of performing large-strain monotonic shearing tests was to evaluate residual strength, which was expected to be relatively low; membrane elastic properties were anticipated to contribute a significant portion of resistance to torsion (particularly at large strains). Corrections for torque developed in confining membranes has previously been investigated for both solid and hollow torsional shear test specimens (e.g., Tatsuoka, et al. 1986 and Frost 1989).

Membrane torque corrections were calculated as a function of θ in monotonic, large strain tests using elasticity theory; a constant membrane thickness and shear modulus were assumed for the membrane material. Torque thus calculated to be contributed by each membrane was then subtracted from the total torque measured at each point during torsional shear tests with the Instron™ axial/torsional loader. Corrections calculated in this manner were approximately 0.05 psi (0.3 kPa) for every one percent shear strain. Figure 9 shows both uncorrected and corrected shear stress versus shear strain data from one of the virgin monotonic torsional shear tests.

        Results. Clean sand specimens exhibited dilative response in all but one post-cyclic monotonic test, wherein the post-consolidation relative density was determined to be about 32.6%. All of the mixtures of fine sand with silt or clay or both exhibited a peak strength at some relatively low shear strain (typically less than 1%), beyond which shear resistance, corrected for membrane torque as described above, decreased to a much lower constant value. In the case of F43 soil, consisting of 80% fine sand with 14% silt and 6% clay, this residual strength value was the lowest measured for all mixtures, namely about 1 psi (6.9 kPa). Observations of distress caused to the Upper San Fernando Dam by the 1971 earthquake imply that shear strains of about 15% may be necessary to develop residual strengths in silty sands (Marcuson, Hynes and Franklin 1992, after Seed, et al. 1973). Residual strengths were estimated for monotonic hollow cylinder torsional shear tests in the present study at a shear strain of 15% for comparison with published residual strength values.

        Peak undrained monotonic torsional shear strengths are compared to residual strengths determined at 15% shear strain in Table 2 for the four soil mixtures tested in this study. The soils were considered to be cohesionless. The reduction in shear strength from peak to residual is significantly higher in the specimens containing 20% fines (F43 and F46) than in the mixture containing 45% fines (F64).

Conclusions

        The results of the laboratory cyclic triaxial and hollow cylinder torsional simple shear test programs summarized in this paper indicate that

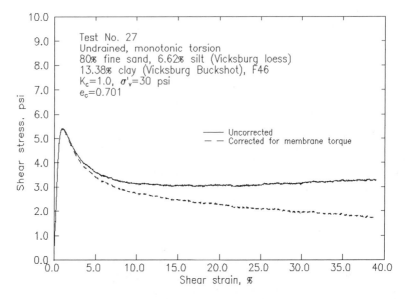

Figure 9.   Shear stress versus shear strain, virgin monotonic torsional
simple shear test on F46 soil specimen

Table 2.   Comparison of peak- to residual undrained shear
strengths in monotonic torsional shear tests on virgin specimens

Soil type	ϕ'[1]	$S_u(peak)$[2], psf (kPa)	S_{ur}[3], psf (kPa)	reduction[4],
F11	35°	(dilative)	(N/A)	(N/A)
F43	33°	2712 (129.9)	158 (7.6)	94%
F46	28.8°	2296 (109.9)	360 (17.2)	84%
F64	26.6°	2019 (96.7)	562 (26.9)	72%

[1]Effective angle of internal friction, determined from effective stress paths
[2]$\sigma'_m \tan\phi'$, where $\sigma'_m = \sigma'_v$ in these isotropically consolidated tests
[3]membrane-corrected strength at 15% shear strain
[4]$(S_u(peak)-S_{ur})/S_u(peak) \times 100\%$

cyclic strength of soils may not be characterized on the basis of gradation alone. The presence of soil particles finer than 0.74 mm (i.e., passing the US #200 standard sieve) does not necessarily assure increased resistance to development of residual excess pore water pressures leading to liquefaction. In fact, for a given global void ratio, it was found from a comprehensive matrix of cyclic triaxial tests on mixtures of sand, silt and clay that cyclic strength was actually reduced to a lower-bound value by the progressive addition of silt- and clay-sized particles until their fractional weight reached 24% to 30% of the total specimen dry weight. Beyond this fractional weight, cyclic strength was found to increase with the addition of fines. It is concluded that sand mixtures containing fines up to about 24% of their dry weight may be inherently collapsible (due, possibly, to the relative compressibility of the finer soil between the sand grains). When fines content exceeds that associated with lower-bound cyclic strength, the fines fraction dominates the cyclic loading response of the soil.

The effects of both gradation and index properties of fines on cyclic strength were investigated. Plasticity index (in the case of this laboratory program, the plasticity index of the fines fraction) exerts much less effect on cyclic strength of soils containing fines at a given void ratio than does the fines content.

Liquefaction is inevitable when residual excess pore pressures reach 70% of the total confining stress in clean sands. Liquefaction of the fine-grained mixtures tested in this study became inevitable when cyclic loading produced residual excess pore pressure ratios of 80% or higher. Strain development accelerated more quickly in mixture soils than in clean sands once liquefaction was inevitable.

Residual strengths of specimens containing 20% low-plasticity fines were found to be very low; deformation potential in such soils would be essentially unlimited. Loose sandy soils containing less than about 30% fines of any plasticity should be considered to have negligible residual strength when liquefied, unless tests on the material prove otherwise.

Acknowledgements

The tests described and the resulting data presented herein, unless otherwise noted, were obtained from research conducted under the Civil Works Investigative Study program (Work Unit 32255 - Liquefaction Potential of Fine-Grained Soils) of the United States Army Corps of Engineers. Permission was granted by the Chief of Engineers to publish this information.

References

Chang, N. Y. (1990). "Influence of fines content and plasticity on earthquake-induced soil liquefaction," Contract Report to US Army Engineer Waterways Experiment Station, Vicksburg, MS, Contract No. DACW3988-C-0078.

Chen, J. W. (1988). "Stress path effect on static and cyclic behavior of Monterey No. 0/30 sand." Ph. D. thesis, University of Colorado, Boulder, CO.

Frost, J. D. (1989). "Studies of the monotonic and cyclic behavior of sands," PhD Thesis, Purdue University, West Lafayette, IN.

Koester, J. P. (1992). "Cyclic strength and pore pressure generation characteristics of fine-grained soils," Ph. D. thesis, University of Colorado, Boulder, CO.

Marcuson, W. F. III, Hynes, M. E., and Franklin, A. G. (1990). "Evaluation and use of residual strength in seismic safety analysis of embankments," Earthquake Spectra, 6(3), 529-572.

Martin, G. R., Finn, W. D. L., and Seed, H. B. (1978). "Effects of system compliance on liquefaction tests," Journal of the Geotechnical Engineering Division, ASCE, 104(GT4), 463-497.

Seed, H. B., Lee, K.L., Idriss, I.M., and Makdisi, F. I. (1973). "Analysis of the Slides in the San Fernando Dams During the Earthquake of February 9, 1971," Report No. EERC-73-2, College of Engineering, University of California, Berkeley, CA.

Steedman, S. and Schofield, A. N. (1991). Personal communication, US Army Engineer Waterways Experiment Station, Vicksburg, MS, 4 April.

Tatsuoka, F., Sonoda, S., Hara, K., Fukushima, S., and Pradhan, T. B. S. (1986). "Failure and deformation of sand in torsional shear," Soils and Foundations, Japanese Society of Soil Mechanics and Foundation Engineering, 26(4), 79-97.

Wang, W. S. (1979). "Some findings in soil liquefaction," Research Institute of Water Conservancy and Hydroelectric Power, Beijing, PRC.

Wang, W. S. (1981). "Foundation problems in aseismatic design of hydraulic structures." In *Proceedings of the Joint US - PRC Microzonation Workshop*, 11-16 September, Harbin, PRC.

Zhu, R. and Law, K. T. (1988). "Liquefaction potential of silt," *Proceedings*, Ninth World Conference on Earthquake Engineering, Tokyo-Kyoto, Japan, August, Vol. III, 237-242.

CYCLIC BEHAVIOR OF PARTIALLY SATURATED COLLAPSIBLE SOILS SUBJECTED TO WATER PERMEATION

Kenji Ishihara*

Kenji Harada**

ABSTRACT

Two series of hydraulically induced collapsibility tests were first conducted, using an oedometer test device, one on a decomposed granite called Masa, and the other on a volcanic soil called Shirasu. A significant amount of volume decrease as large as 14 % was observed in the test specimens as a result of hydraulic collapse due to infiltration of water. The test results also revealed that wetting-induced volume change tends to increase with decreasing density and water content at which a sample is prepared by compaction. On the basis of the test results as above, partially saturated samples with different histories of collapse were prepared and tested under cyclic loading conditions using a simple shear test apparatus. It was discovered that cyclic resistance of the test samples having experienced larger collapse was smaller as compared to that of the samples having undergone a smaller degree of hydraulic collapse, even though the density and water content of the samples were almost identical. In addition, it was shown that the cyclic resistance of collapsible soils becomes notably small, even when they are partially saturated. It may thus be mentioned that fines-containing sands with high potential to hydraulic collapse are in an unstable condition susceptible to liquefaction or cyclic softening due to dynamic disturbance such as seismic shaking, if they are in a state wetted by water invasion.

* Professor of Civil Engineering, University of Tokyo, 7-3-1 Hongo, Bunkyo-ku, Tokyo, 113 JAPAN
** Civil Engineer, Fudo Construction Co., 1-2-1 Taitow, Taitow-ku, Tokyo, 110 JAPAN

INTRODUCTION

Collapsing soils have been recognized as being of engineering significance in connection with settlements of structures constructed on windlaid loess deposits as well as with subsidences of very thick fills and high embankments. It is generally recognized that the volume decrease due to collapse by water permeation can potentially take place in sandy or silty soils containing some percentage of clay fraction, if they are deposited or placed in a loose state with a low water content. In natural environments, loess deposits with aeolian origin have been known as a typical type of collapsible soils and many studies have been made thus far to identify subsidence characteristics of these soils (Clevenger, 1958; Dudley, 1970, Chen et al. 1987, Amirsoleymani, 1989). Soils compacted dry of optimum to low density with low water content have been known as another type of soils exhibiting high vulnerability to hydraulic collapse. Many studies have been made as well to investigate the mechanism and factors influencing the collapse of compacted soils (Cox, 1978, Lawton et al. 1989).

In most of the studies as above, attention has been drawn primarily to the adverse effects such as volume change or settlements characteristics of soils resulting from collapse due to water permeation. However, a recent experience of a series of catastrophic landslides due to an earthquake which occurred in the loess deposit in Soviet Tajik appears to have given rise to needs for investigating the collapsibility of soils in relation to shear strength under dynamic loading conditions (Ishihara et al. 1990). In fact, the major cause for the destructive landslide was identified as liquefaction of the windlaid silty loess deposit which had been wetted over the years by permeation of irrigation water. The loess deposit which had been on the verge of shear failure due to the collapse by water invasion was further subjected to a shaking during the earthquake and easily led to the extensive liquefaction and consequent mud flow. Thus, the combined action of hydraulic collapse and seismic shaking was the major cause for the catastrophe in the Tajikistan Republic of USSR.

On the basis of the observation as above, it was considered necessary to obtain some basic test data regarding the behavior of collapsible soils subjected to a superimposed action of hydraulic collapse and cyclic loading. In this context, multiple series of cyclic simple shear tests were conducted on artificially prepared samples of decomposed granite and volcanic soil which have undergone more or less collapse by water permeation. The conduct of the tests and its consequence are described in the following pages.

SAMPLE PREPARATION

The materials used for artificially preparing collapsible samples are

silty sands from weathered granite and volcanic deposit. The grading of
these material is shown in Fig. 1.

(1) Collapsibility tests

        To determine the collapsible characteristics of the materials,
oedometer tests were carried out. The samples were prepared by
compacting the material with different water contents in the oedometer test
ring. Disk-shaped specimens 2 cm thick and 6 cm in diameter were
prepared by tamping the material in two layers with a brass-made flat-
bottom rammer. Then, a vertical overburden pressure of σ'_{v0} = 98kPa was
applied to the specimen, before it was inundated with water. Upon water
infiltration, the partly saturated specimen underwent collapse in its
structure accompanied by reduction in its volume. After the wetted
specimen had stopped shrinking thereby coming to an equilibrium water
content, the decrease in the specimen thickness, h, was measured. The
collapsibility coefficient, η, was defined as,

$$\eta = \frac{\Delta h}{h_0} = \frac{\Delta e}{1 + e_0} \qquad \text{..... (1)}$$

where h_0 and e_0 denote, respectively, the initial height and initial void ratio
of the specimen before soaking. Δe and Δh are the decrease in void ratio and
height of the specimen, respectively, due to soaking.

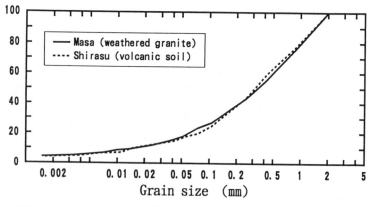

Fig. 1 Grain size distribution of the test materials

(2) Cyclic Simple Shear test

The test apparatus used in this study is the same as that employed in the previous studies (Ishihara and Yamazaki, 1980; Nagase and Ishihara, 1987). This apparatus consists of three main components : two mutually perpendicular horizontal loading devices, an assemblage for specimen placement, and an equipment for applying vertical load. A schematic illustration of the apparatus is shown in Fig. 2. Although the test apparatus is capable of applying simple shear stress independently in two directions, only one component of loading device was used in this study. Therefore, all the tests were the conventional type of uni-directional simple shear loading test. Samples 3 cm thick and 7 cm in diameter were prepared in a membrane-enclosed space which is surrounded by stacks of annular plates in the shear test apparatus. The specimen with a pre-determined moisture content was placed and compacted by means of a flat-bottom rammer 5 cm in diameter.

After the partially saturated compacted specimen is prepared, a vertical overburden stress of $\sigma'_{v0} = 98$ kPa was applied and then water was permeated through the specimen. The volume of the specimen starts to decrease due to collapse in the structure but it reached a stationary state within 2 hours. Then the decrease in thickness of the specimen was

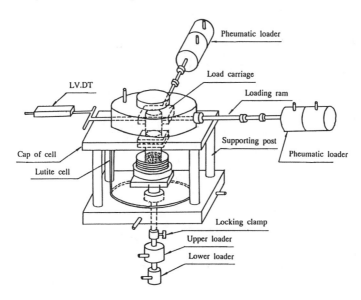

Fig. 2 Simple shear test apparatus

measured to determine the collapsibility coefficient defined by Eq. (1).
Cyclic simple shear stress with constant amplitude was then applied until
the specimen softened producing ± 3.5 % amplitude of simple shear strains.
During the cyclic load application, thickness of the specimen was
maintained unchanged by preventing vertical movement of the specimen's
cap. Therefore, the cyclic loading was conducted under constant volume
conditions.

Since the specimen is enclosed by stacks of annular plates, K_0-
conditions appear to be realized when the vertical stress is applied and also
in the process of volume decrease due to hydraulic collapse. The specimen
was not fully saturated even after infiltrating water and, therefore, pore
water pressure was not monitored during the cyclic stress application.

TEST RESULTS

(1) Hydraulic Collapsibility Characteristics

     In order to identify the compaction characteristics of the material

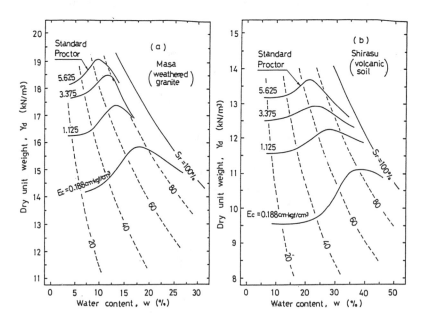

Fig. 3 Compaction curves of the test material

used in the tests, a series of compaction tests was carried out following the procedures specified in the Standard Proctor Tests. The family of compaction curves employing different compacting energy are presented in Fig. 3. The curve yielding the highest density was obtained by what is called the Standard Proctor Test in which a compacting energy employed was $Ec = 5.625$ cm $\cdot$ kgf/cm^2. The other curves are those obtained by imparting lesser amount of compacting energy. It may be seen in Fig. 3 that the line of optimums connecting coordinates of optimum water content and maximum dry density does coincide approximately with the contour line of saturation of 70 % for the silty sands used in the present study.

One of the results of collapsibility tests using the oedometer ring is demonstrated in Fig. 4 in which specimens prepared at identical initial dry unit weight of $\gamma_{d0} = 14.2$ and 10.3 kN/m^3 but with varying water contents were loaded and subsequently inunated with water. It can be seen in Fig. 4 that, while there is negligible change in the specimen height under a sustained vertical stress of $\sigma_{v0} = 98$ kPa, a sudden decrease in the specimen height took place as a consequence of hydraulic collapse due to water inundation. It is also seen that the amount of wetting-caused volume change tends to significantly increase with decreasing water content at which the sample has been compacted.

The results of the test series using the oedometer device to examine the effects of water content at compaction under a given density are presented in Fig. 5. It is to be noted that different compacting energy was employed to prepare specimens with different water contents while maintaining a constant density. Fig. 5 shows that the amount of wetting-induced collapse under a given vertical overburden stress depends mainly on the dry unit weight and water content at which the soil has been compacted. It is apparent that the drier and the looser the soil was at compaction, the greater was the amount of collapse due to water inundation, as represented by the collapsibility coefficient, η. It is also noted in Fig. 5 that if the specimen is compacted at a dry density greater than about 15.7 and 11.2 kN/m^3 for Masa and Shirasu, respectively, practically no collapse can take place.

The same test data as those shown in Fig. 5 are presented in Fig. 6 in terms of the collapsibility coefficient versus the saturation ratio, Sr, at compaction. It is to be noted that there is a relation between water content, w, and saturation ratio, Sr, as follows,

$$S_r = \frac{w}{\gamma_w / \gamma_d - 1/G_s} \qquad \ldots (2)$$

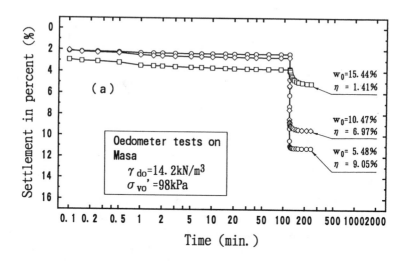

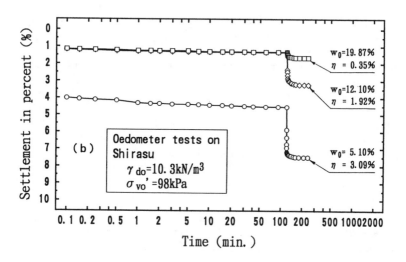

Fig. 4 Collapsibility tests using an oedometer device

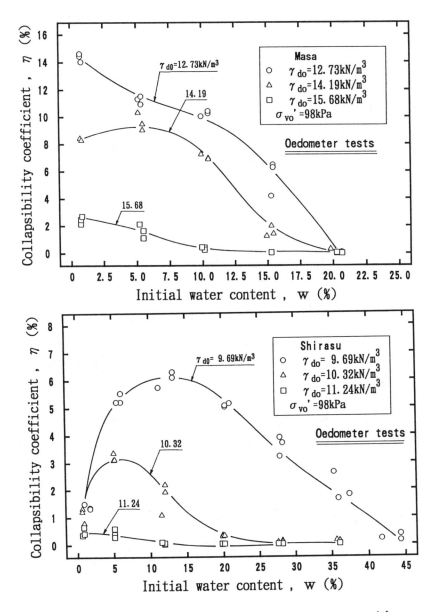

Fig. 5 Collapsibility versus initial water content for the test material

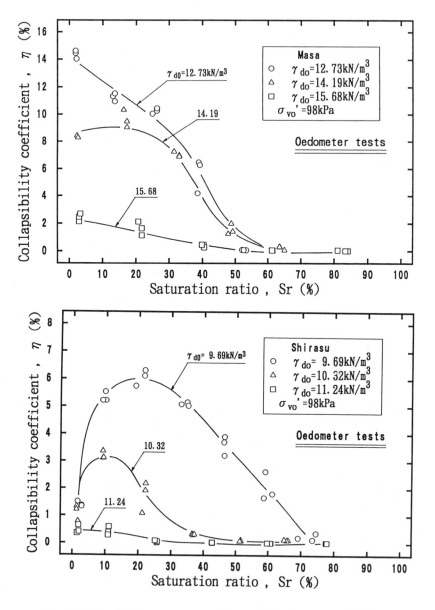

Fig. 6 Collapsibility versus saturation ratio for the test material

where γ_w is unit weight of water and Gs is specific gravity of solid particle contained in soils. Therefore, given a value of dry density, γ_d, the saturation ratio can be computed by Eq. (2) for a known water content.

The saturation ratio plotted in the abscissa of Fig. 6 was obtained in this way from the water content values shown in Fig. 5. Fig. 6 indicates the characteristic relationship among the collapsibility coefficient, saturation ratio and the as-compacted density of the material tested under the effective overburden pressure of $\sigma'_{v0} = 98$ kPa. It is apparent that the looser the material compacted and the smaller the saturation ratio, the greater was the soaking-induced volume decrease and hence the collapsibility. The results of the hydraulic collapsibility tests shown in Fig. 5 can be displayed alternatively in a form of an isogram indicating equal collapsibility coefficient in the diagram of dry unit weight versus water content. In fact, in the plot of Fig. 5 it is possible to read off combinations of values of water content and dry unit weight which yield the same collapsibility coefficient, η. Pairs of such data are laid down in a diagram of Fig. 7 to draw up a

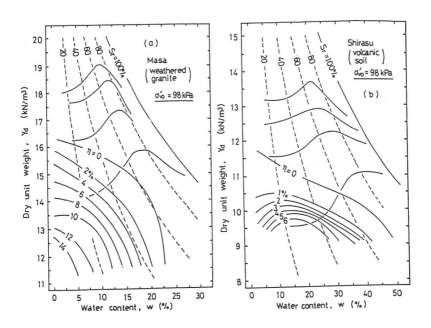

Fig. 7 Isogram for equal collapsibility

family of contour lines giving equal values of collapsibility coefficient. Lawton et al. (1989) indicated that there exists an upper limit in the saturation ratio under a given overburden pressure, beyond which the hydraulic collapse can not occur. In the Masa and Shirasu used in the present study, this upper limit is shown in Fig. 7 to be about Sr = 60 -70 % under the overburden pressure of 98 kPa. It was also noted by Lawton et al (1989) that, for a given overburden pressure, there exists a critical dry unit weight for which no volume change whatsoever can occur in a soil upon wetting regardless of the molding water content. This upper bound in the critical unit weight for the materials tested appears to be about 16 kN / m³ for Masa and 11.0 kN / m³ for Shirasu, as accordingly indicated in Fig. 7. It should be noted that this critical unit weight tends to increase significantly with the overburden pressure, although this aspect is not addressed in this paper.

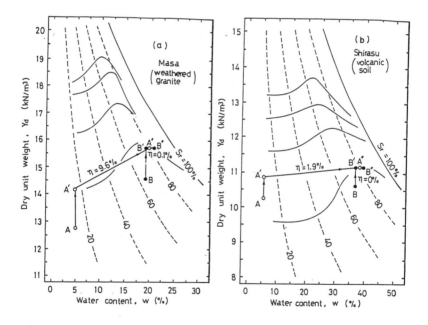

Fig. 8 Changes in the states of samples during application of
overburden pressure and water permeation

(2) Cyclic Simple Shear Tests

Specimens in the mold of the simple shear test were prepared at different water contents and dry unit weights, but upon being subjected to an overburden pressure and upon permeation by water, the specimens reduced their volume, accompanied by an increase in water content, reaching an almost identical condition at the end. This change in the state of the sample is illustrated by arrows in Fig. 8. For example, point A in the figure indicates an initial state of a loose sample of Masa compacted to an initial density of γ_{d0} = 12.7 kN/m^3 at an initial water content of ω_0 = 5 %. When the overburden pressure of σ'_{v0} = 98 kPa was applied to the sample, it increased the density to a value of 14.1 kN/m^{3,} as indicated by point A' in Fig. 8. When the sample was subsequently infiltrated with water, its water content increased to a value of ω = 20 %, accompanied by a further increase of dry density to a value of γ_d = 15.6 kN/m^3, as indicated by point A" in Fig. 8. During this water-absorbing process, a hydraulic collapse as high as η = 9.6 % was observed. Similar process of change of state for another sample compacted denser initially is indicated by point B to B' and to B" in Fig. 8, whereby the collapse was as small as η = 0.1 %. It is to be noted that, while the initial states were significantly different between these two samples, the final state reached after wetting was practically identical with the dry density of 15.6 kN/m^3 and the water content of about 20 %. The corresponding saturation ratio was about 80 % which is slightly higher than the value corresponding to the optimum water content. This final state of the samples also coincides with the occlusion degree of saturation where air is occluded in the pore water and migrated only with pore water.

The cyclic loading phase of the tests was executed after the samples had reached the identical state of dry density and water content as mentioned above. The results of the cyclic simple shear tests conducted under constant-volume conditions are presented in Fig. 9 where the cyclic stress ratio required to cause ± 3.5 % shear strain is plotted versus the number of cycles. The cyclic stress ratio is defined simply as the amplitude of uniform simple shear stress, τ_d, normalized to the overburden pressure, σ'_{v0}, which has been applied to the sample since it was consolidated. As indicated in the inset of Fig. 9, the samples just prior to cyclic load application had been brought to approximately the same conditions with respect to the density, water content and saturation ratio. The only difference was the process or history in hydraulic collapse which the samples had undergone before they reached the same final state. This difference in the process is represented by the collapsibility coefficient, η. It is apparent in Fig. 9 that the cyclic stress ratio great enough to produce a state of cyclic softening with ± 3.5 % shear strain under a given number of

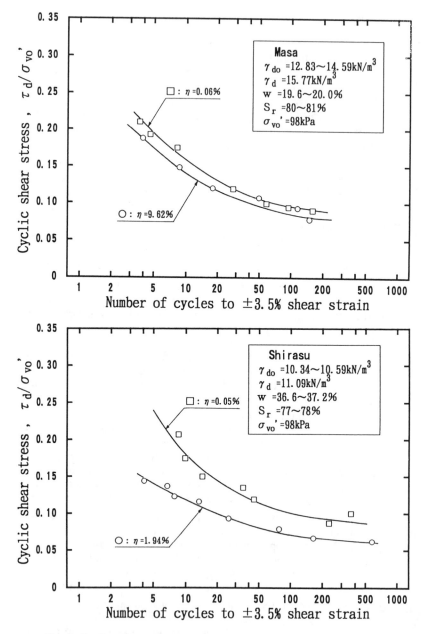

Fig. 9 Cyclic stress ratio versus number of cycles

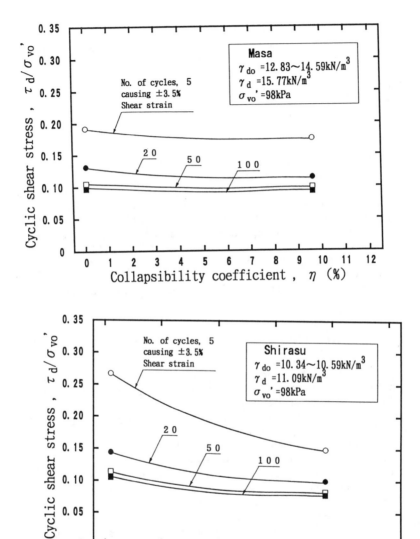

Fig. 10 Cyclic strength versus collapsibility coefficient

cycles depends significantly on the degree of hydraulic collapse experienced by the samples prior to the cyclic load application. The greater the hydraulic collapse, the smaller the cyclic stress ratio required to cause cyclic softening. In order to see this effect, the cyclic stress ratio needed to induce ± 3.5 % shear strain under the number of cycles of 5, 20, 50 and 100 was read off from Fig. 9 and plotted in Fig. 10 versus the collapsibility coefficient. It can be clearly seen in Fig. 10 that, at any given number of cycles, the cyclic stress ratio required to induce cyclic softening in the material tested tends to decrease as the collapsibility coefficient increases.

No less important than this would, however, be the smallness of the cyclic strength itself, as witnessed by a value on the order of 0.15 of the cyclic stress ratio causing ± 3.5% shear strain. It is to be noticed that the above magnitude is as low as the cyclic strength observed in fully saturated loose sand (Ishihara, 1993). On the other hand, it has been recognized that the effects of saturation on liquefaction resistance act towards increasing the cyclic strength with decreasing saturation ratio (Chaney, 1978). However, the above conclusion by Chaney (1978) appears to be valid only for water-sedimented or non-collapsible soils. If soils are in a collapsible state, the cyclic strength becomes significantly low even when the saturation ratio is 60 to 80 %.

CONCLUSIONS

In order to examine the influence of hydraulic collapse on the liquefaction or cyclic softening characteristics of sandy soils, two series of cyclic simple shear tests were conducted on the Japanese local soils. Before conducting this cyclic simple shear test, another series of basic tests for identifying the collapse characteristics was also performed using the normal type of an oedometer test device. This test consists in applying an overburden pressure of 98 kPa and then infiltrating water through the specimens initially prepared to different dry densities by compaction with different water content. The amount of volume decrease per unit volume caused by wetting is termed "collapsibility coefficient" and used as an index parameter to identify the collapse characteristics. The outcome of these tests using the oedometer device indicated that the collapsibility coefficient as much as 14 % is obtained for the materials tested, if they are prepared by compaction at a dry density and water content far dry of Proctor optimum. It was also shown that the collapsibility tends to decrease with increasing density and water content at compaction, until it eventually becomes equal to zero with an increased saturation ratio which takes a limiting value around 60 - 70 %.

In the cyclic simple shear tests, samples were prepared to different densities at different water contents, but in such a way that all the samples would change to possess a nearly identical density and water content after the hydraulic collapse had taken place upon being wetted by water infiltration. The cyclic loading tests were thus performed on samples having an almost identical density and water content, but with different degree of collapsibility experienced in the preceding process of wetting. The outcome of the cyclic simple shear tests conducted under constant-volume conditions disclosed that the cyclic strength is as small as the one observed in fully saturated loose sand and that, with increasing degree of hydraulic collapsibility, the cyclic strength of fines-containing sand can be significantly reduced. Thus, the collapsible soils were found to exhibit low resistance to liquefaction, even though they are in a partially saturated state. The deteriorating effects of collapsibility, combined with seismically-induced cyclic loading, would indicate potentially unstable nature of the water-infiltrated loess deposit as witnessed in Soviet Tajik area which have developed extensive liquefaction and mud flow during the recent earthquake (Ishihara et al. 1990).

ACKNOWLEDGEMENT

The investigation described in this paper was conducted under the sponsorship of the grant-in-aid from the Ministry of Education of the Japanese Government.

REFERENCES

Amirsoleymani, T. (1989), "Mathematical Approach to Evaluate the Behavior of Collapsible Soils," Proc. 12th International Conference on Soil Mechanics and Foundation Engineering, Rio de Janeiro, Vol. 1, PP. 575-582.

Chang, R.C. (1978), "Saturation Effects on the Cyclic Strength of Sands", Earthquake Engineering and Soil Dynamics, ASCE, Vol.1 PP. 342-358.

Chen, Z.Y., Qian, H.J., and Bao, C.G. (1987), "Problems of Regional Soils," Proc. 8th Asian Regional Conference Soil Mechanics and Foundation Engineering, Kyoto, Japan, Vol. 2, PP. 167-190.

Clevenger, W.A. (1958), "Experiences with Loess as Foundation Material," Transaction of ASCE, No. 2916, PP. 151-169.

Cox, D.W. (1978), "Volume Change of Compacted Clay Fill," Proc. Conference on Clay Fills, Institution of Civil Engineers, London, PP. 79-87.

Dudley, J.H. (1970), "Review of Collapsing Soils," Journal of ASCE, SM3, PP. 925-947.

Holtz, W.G. (1948), "The Determination of Limits for the Control of Placement Moisture in High Rolled Earth Dams," Prod. ASTM, Philadelphia, Pa., PP. 1240-1248.

Ishihara, K. and Yamazaki, F. (1980), "Cyclic Simple Shear Tests on Saturated Sand in Multi-Directional Loading," Soils and Foundations, Vol. 20, No. 1, PP. 45-59.

Ishihara, K., Okusa, S., Oyagi, N. and Ischuk, A. (1990), "Liquefaction-Induced Flow Slide in the Collapsible Loess Deposit in Soviet Tajik," to appear in Soils and Foundations.

Ishihara, K. (1993), "Liquefaction and Flow Failure during Earthquakes", Geotechnique, No.3, Vol.43, PP. 351-415.

Jennings, J.E. and Knight, K. (1957), "The Additional Settlement of Foundations due to a Collapse of Structure of Sandy Subsoils on Wetting," Proc. 4th International Conference on Soil Mechanics and Foundation Engineering, London, Vol. 1, PP. 316-319.

Jennings, J.E. and Knight, K. (1975), "A Guide to Construction on or with Materials Exhibiting Additional Settlement due to Collapse of Grain Structure," Proc. 6th Regional Conference for Africa on Soil Mechanics and Foundation Engineering, Duban, South Africa, PP. 99-105.

Lawton, E.C., Fragaszy, R.J. and Hardcastle, J.H. (1989), "Collapse of Compacted Clayey Sand," ASCE, Journal of Geotechnical Engineering, Vol. 115, No. 9, PP. 1252-1267.

Nagase, H. and Ishihara, K. (1987), "Effects of Load Irregularity on the Cyclic Behavior of Sand," Soil Dynamics and Earthquake Engineering, Vol. 6, No. 4, PP. 239-249.

LIQUEFACTION IN SILTY SOILS: DESIGN AND ANALYSIS

W.D. Liam Finn[1] R.H. Ledbetter[2] and Guoxi Wu[1]

ABSTRACT

The conditions for triggering liquefaction in silty soils and assessing the consequences of liquefaction are reviewed. Recent findings on the evaluation of residual strength and the effects of effective stress on the resistance to liquefaction are presented. The major impact of these findings on the potential costs of remediation are demonstrated. The procedures for estimating the consequences of liquefaction and for evaluating the effectiveness of remedial measures are described by means of case histories.

INTRODUCTION

The evolution of the state-of-the-art for evaluating the seismic safety of earth structures such as embankment dams which contain potentially liquefiable soils in the embankment or foundation is reviewed and the impact of recent research findings and field studies is described. There are three fundamental questions involved in such seismic safety evaluations: (1) Will liquefaction be triggered?; (2) What are the expected consequences of liquefaction?; and (3) What remedial measures are practical and cost-effective? In this context, liquefaction means the collapse of contractive soils under undrained conditions

---

[1] Department of Civil Engineering, University of British Columbia, Vancouver, B.C., Canada

[2] US Army Corps of Engineers, Waterways Experiment Station, Vicksburg, Mississippi, U.S.A.

with significant reduction in shearing resistance from peak undrained strength to residual or steady state strength (Fig. 1). If the existing static shear stresses in the dam exceed the residual strength, the potential exists for large deformations.

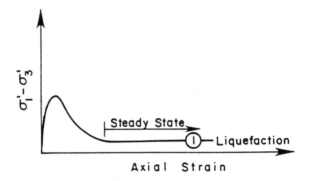

Fig. 1. Undrained behaviour of contractive sands.

Ideally, investigations of the triggering of liquefaction should be conducted on undisturbed samples from the field. However, because of the difficulty of obtaining undisturbed samples in such contractive materials in the current state of practice, in-situ methods such as the standard penetration test, cone penetration test and shear wave velocity measurements, are used to define the triggering of liquefaction. These methods rely on empirical correlations between measures of penetration resistance and the performance of soil deposits in past earthquakes. The database on triggering is based primarily on data from level ground and at pressures less than approximately 100 kPa (1 ton/sq.ft.). To extend the database to depths representative of foundation soils under dams requires the use of correction factors for the effects of overburden pressure and initial static shear stress. At the high overburden pressures typical of dam foundations, the use of these factors results in large reductions in initial estimates of liquefaction resistance. Recent laboratory research on reconstituted samples and on frozen samples from the field have shown that generalized correction factors in current use may be overly conservative and can lead to excessive remediation costs in some cases.

How the consequences of liquefaction and the effects of remediation are evaluated also has a major impact on the cost of remediation. Until very recently, these evaluations were made primarily on the basis of achieving an acceptable factor of safety typically in the range of 1.1 to 1.2. In the last few years there has been increasing emphasis on evaluating the consequences of liquefaction by estimating the post-liquefaction deformations and using deformation criteria to control the location and extent of remediation measures. In a number of cases this approach has resulted in considerable savings in the cost of remediation. The approach based on deformations will be illustrated by two case histories from engineering practice. One example will demonstrate the calculation of large deformations in a tailings dam, and the other will illustrate the evaluation of remediation measures to stabilize the upstream slope of a water retaining dam.

TRIGGERING OF LIQUEFACTION

The triggering of liquefaction is evaluated most often in practice using the liquefaction assessment chart developed by Seed et al. (1985) and shown in Fig. 2. The chart shows the boundary line between liquefiable and nonliquefiable level sandy sites with various percentages of fines in the sands for an earthquake of magnitude 7.5. The effects of the earthquake are characterized by the cyclic stress ratio τ/σ'_{vo} where τ is the effective uniform cyclic shear stress during the earthquake, and σ'_{vo} is the effective overburden pressure. The resistance of the site to liquefaction is specified by the standard penetration resistance, N, normalized to a vertical effective overburden pressure of 100 kPa and an energy level of 60% of the free-fall energy of the hammer. This resistance is designated $(N_1)_{60}$. The critical liquefaction resistances for earthquakes of other magnitudes are obtained from the standard case of $M = 7.5$ using the correction factors in Table 1.

Table 1: Correction Factors for Magnitude (Seed and Harder, 1990)

Earthquake Magnitude (M)	Correction Factor for Magnitude
8.50	0.89
7.50	1.0
6.75	1.13
6.00	1.32
5.25	1.5

K_σ - Correction for Overburden Pressure

For pressures greater than 100 kPa, the liquefaction resistance from the chart is reduced by using a correction factor K_σ. A representative relationship

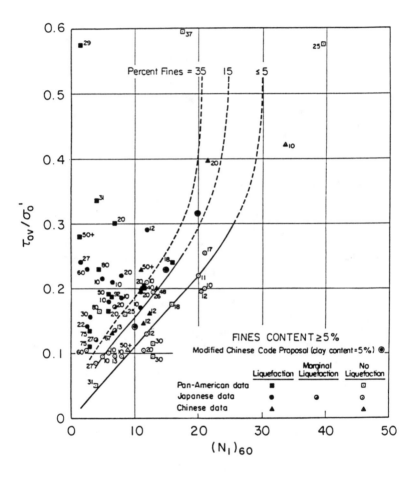

Fig. 2. Relationship between stress ratio causing liquefaction and $(N_1)_{60}$ values for silty sand for M=7.5 (Seed et al., 1985).

between K_σ and σ'_{vo} recommended by Seed and Harder (1990) is shown in Fig. 3. This curve is based on data from a wide variety of sand types at various relative densities which was summarized by Harder (1988). There is wide scatter in the data. It may be seen that at pressures of 800 kPa, this factor reduces the liquefaction resistance to 45% of its original value. Clearly the correction factor

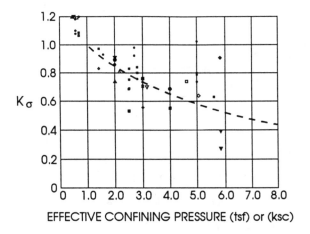

Fig. 3. K_σ - Correction for Overburden Pressure (Seed and Harder, 1990).

can have a major impact on the potential for triggering liquefaction and hence on the cost of remediation.

Recent studies have demonstrated that K_σ-σ'_{vo} relationships developed for specific sand types show much less variation and that for some sand types, the corrections to the liquefaction resistances from Fig. 2 are substantially less than suggested by the generalized curve in Fig. 3.

Laboratory studies on reconstituted samples by Vaid and Thomas (1994) on a variety of sands have shown that K_σ for any given sand, is a function of relative density, and that the correction factor increases with increasing relative density. The findings by Vaid and Thomas (1994) are shown in Fig. 4. It is clear that for these sands, the correction factor may be considerably less from that suggested in Fig. 3.

Recently, field studies were conducted in the foundation materials of Duncan Dam in British Columbia as part of an evaluation of the seismic safety of the dam (Pillai and Stewart, 1993). Frozen samples were retrieved from potentially liquefiable sands and tested in the laboratory under confining pressures up to 1200 kPa. No change in liquefaction resistance was noted over the pressure range of 100 kPa to 1200 kPa. Of course in this case, the effects of effective overburden pressure were masked by the decreases in void ratio caused by the

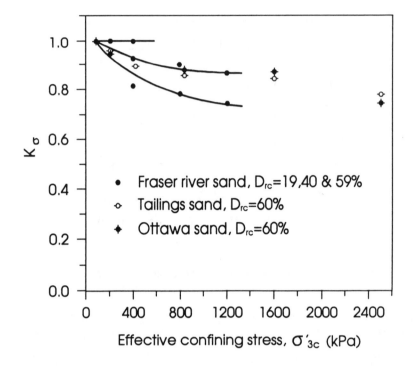

Fig. 4. K_σ - Factors for Clean Sand as a Function of Relative Density and Sand
Type (Vaid and Thomas, 1994).

increasing pressure. In-situ void ratio was evaluated by means of nuclear logging
and when corrections for void ratio changes were made, the liquefaction
resistance of samples of equal void ratio were estimated to be reduced by 40% as
the effective confining pressure varied from 100 kPa to 900 kPa. This results in
a $K_\sigma = 0.6$ compared to $K_\sigma = 0.4$ from the Seed and Harder (1990) correlation.
Obviously the impact of these differences on estimates of dam performance are
very significant. In the case of the Duncan Dam, using the liquefaction resistance
of the frozen samples rather than the liquefaction resistance derived from the
Seed liquefaction resistance chart modified by Seed and Harder (1990) K_σ values
meant that no remediation was considered necessary. Clearly, it is desirable,
where possible, to use site specific values of K_σ.

K_α Correction for Initial Static Shear

A correction factor, K_α, for initial static shear has also been proposed (Seed and Harder, 1990). The role of K_α is quite different for contractive and dilative sands. In the case of contractive sands, liquefaction in the sense of a major reduction in shearing resistance occurs and the presence of initial static shear in these unstable materials facilitates triggering. Therefore the liquefaction resistance is reduced by the presence of the static shear. For dilative materials liquefaction is defined as the development of a specified amount of strain. In this case there is concern with cyclic mobility which is an entirely different physical phenomenon. Generally, in dilative materials, the development of strain is inhibited by the presence of static shear. The database on K_α is too general for effective use in design and site specific values should be determined for this factor also.

RESIDUAL STRENGTH

The ideal situation would be to measure the residual strength of undisturbed samples from the field. This can only be done with confidence on frozen samples and this implies also that such samples can have only a limited amount of fines. Otherwise the soil structure will be disturbed by the freezing process as drainage will not be able to occur during freezing.

Poulos et al. (1985) have developed procedures for correcting for the effects of sample disturbance the residual strengths determined by testing of good quality sand samples retrieved from test pits or by fixed piston sampling. The corrections are very sensitive to the slope of the steady state line. This procedure has been discussed in more detail by Finn (1990).

Seed (1987), on the basis of a study of case histories, suggested a correlation between residual strength and liquefaction resistance. An updated version of this correlation is shown in Fig. 5 (Seed and Harder, 1990). Typically, in practice, the lower bound curve is used with the median or 33 percentile of the penetration resistances $(N_1)_{60}$. This procedure results in very low residual strengths for many projects in which $(N_1)_{60}$ is less than 15. A re-evaluation of the same data suggested that the residual strength might be affected by the effective overburden pressure (Lo et al., 1991; McLeod et al., 1991). In the seismic safety assessment of Sardis Dam (Finn et al., 1991), the residual strength of a clay silt was taken as $0.075 \, \sigma'_{vo}$, a value based primarily on in-situ vane tests.

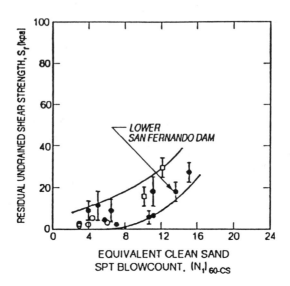

Fig. 5.   Correlation Between Residual Strength and $(N_1)_{60}$ (Seed and Harder, 1990).

Vaid and Thomas (1994) conducted a laboratory study on reconstituted samples of sand to evaluate the effect of overburden pressure on residual strength using extension tests. Their findings are shown in Fig. 6. Note the wide range in residual strengths with effective confining pressure at any given void ratio. For this particular data, the residual strength varies between 0.1 σ'_{vo} and 0.18 σ'_{vo}. Data from Vaid et al. (1989) suggests that the residual strength is also stress path dependent. This implies that different residual strengths would apply along different portions of a sliding surface.

The residual strengths determined from the frozen samples in the Duncan Dam study, showed that the residual strength was of the order of 0.21 σ'_{vo} (Byrne et al., 1993). This is similar to the relationship between undrained strength and effective overburden pressure for normally consolidated clays. These strengths are similar to the mean strengths suggested by Seed and Harder (1990) for effective overburden pressures around 100 kPa, but the differences between the two estimates continue to diverge significantly with increasing

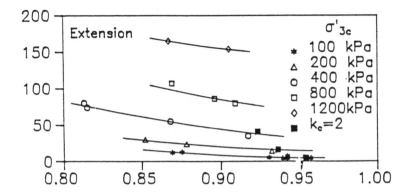

Fig. 6.   Dependence of Residual Strength on Effective Overburden Pressure (Vaid and Thomas, 1994).

effective confining pressure.   In the case of the Duncan Dam, the mean Seed and Harder (1990) strength was only one-third of the measured strength at an overburden pressure of 500 kPa (Byrne et al., 1993).

Impact of Recent Developments

The two major parameters controlling seismic safety and the cost of remediation are the resistance to the triggering of liquefaction and the residual strength.   Laboratory research findings and the studies conducted for the Duncan Dam suggest that current practice may be very conservative in some instances. Estimates of the correction to liquefaction resistance for effective confining pressure may be too high in some cases and estimates of residual strengths at the high effective confining pressures typical of dam foundations may be too low. The combined effect of these two factors in such cases, is to inflate significantly the cost of remediation.   Therefore, there is a strong economic incentive to undertake a site-specific fundamental study of the potential for triggering liquefaction and of the level of available residual strength as was done in the Duncan dam study.

EFFECTS OF FINES CONTENT ON LIQUEFACTION RESISTANCE

Non-Plastic Fines

The data in Seed's resistance chart (Fig. 2) shows that the resistance to triggering liquefaction can be substantially increased by the presence of fines. This increase becomes relatively more important at the lower penetration resistances. These findings from field experience have spurred an interest in the effects of fines content on liquefaction resistance and residual strength. Many laboratory studies have been conducted, primarily on reconstituted sand samples in which various percentages of fines have been incorporated. In spite of these studies, there still appears to be considerable confusion about the effect of fines on liquefaction resistance. The confusion arises from the different ways in which comparisons are made of the liquefaction resistances of sands with various percentages of fines.

Seed et al. (1985) reviewed sites that did and did not liquefy during earthquakes where the fines content was greater than 5%. They found that for the same penetration resistance the liquefaction resistance increased with increasing fines content (Fig. 2). The field data in Fig. 2 shows a strong dependency of liquefaction resistance on fines content. In the case of the field data, for the <u>same normalized penetration resistance</u> $(N_1)_{60}$ the liquefaction resistance increases with increasing fines content.

What the soils being compared have in common, in this case, is the $(N_1)_{60}$ value. If the basis for comparison is changed the conclusion may well be different. For example, Troncoso (1990) compared the cyclic strength of tailings sands with different silt contents ranging from 0 to 30% at a <u>constant void ratio</u>. On this basis of comparison he found that cyclic strength <u>decreased</u> with increased silt content (Fig. 7). Kuerbis and Vaid (1989) studied the effects of fines using samples deposited in a manner that replicated field deposition conditions. They found for the sand tested that up to 20% fines could be accommodated within the sand skeleton and that samples with <u>the same sand skeleton void ratio</u> had the <u>same cyclic strength</u> for fines content less than 20%.

The different conclusions about the effects of fines on liquefaction resistance arising from the different comparison criteria are summarized in Table 2.

Another way of looking at the field data is to note that the penetration resistance of silty sands is somewhat lower than for clean sands <u>at the same cyclic strength</u>. Corrections based on the curves in Fig. 2, can be applied to the $(N_1)_{60}$ values measured in the silty sands and then the curve for clean sands may be used

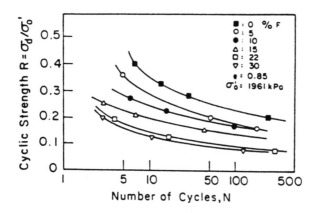

Fig. 7. Variation in cyclic strength with fines content (Troncoso, 1990).

Table 2. Effects of Fines on Cyclic Strength.

Criterion of Equivalency Clean and Silty Sand	Effect of Fines on Cyclic Strength
1. Same $(N_1)_{60}$	Increase
2. Same gross void ratio	Decrease
3. Same void ratio in sand skeleton	No change while fines content can be accommodated in sand voids

to determine cyclic strength for the corrected $(N_1)_{60}$ value. Such corrections have been widely used.

If the fines are plastic, liquefaction criteria based on field data from Chinese earthquakes should be taken into account (Wang, 1979) in determining if the soil will liquefy.

Plastic Fine-Grained Soils

Liquefaction potential of weak clays is usually determined using the Chinese criteria developed by Wang (1979). These criteria are:

- per cent finer than 0.005 mm $\leq 15\%$
- liquid limit, LL $\leq 35\%$
- natural water content ≥ 0.9 LL
- liquidity index, $I_w \leq 0.75$

Soils which satisfy all four criteria are judged vulnerable to liquefaction or significant strength loss. The Chinese criteria are often applied strictly with no account taken of uncertainties in the measurements of the parameters of the criteria. The U.S. Army Corps of Engineers and some major consulting firms now use an error term when applying the criteria.

In connection with remediation studies for Sardis Dam (Finn et al., 1991), the U.S. Army Corps of Engineers reviewed reports on the scatter in measured index properties in their own laboratories over the last 30 years to determine the likely range of variations in test data. In addition tests were also conducted on samples of standard soils of low to medium plasticity in order to establish the scatter in the data from the laboratory of the US Army Corps of Engineers District responsible for Sardis Dam. These standard soils are used to check comparability of testing procedures between different Corps of Engineers' laboratories and private laboratories. As a result of these studies the following changes in measured properties were adopted before applying the Chinese criteria (ignoring the liquidity index):

- decrease the fines content by 5%
- decrease the liquid limit by 2%
- increase the water content by 2%

These changes led to a substantial increase in the extent of liquefiable clayey silt under Sardis Dam. Therefore, an investigation of Chinese procedures was undertaken by Koester (1990). The Chinese determine the liquid limit using a fall cone rather than the Casagrande device generally used in North America. Using a standard Chinese fall cone and following Chinese standard SD 128-007-84, Koester (1990) showed that the fall cone gives a liquid limit about 3% to 4% greater than the Casagrande device. Further studies reported later by Koester (1992) confirm the findings regarding the liquid limit.

On the basis of all the above studies the following changes in measured index properties were finally adopted to account for uncertainty before application of the Chinese criteria:

- decrease the fines content by 5%
- increase the liquid limit by 1%
- increase the water content by 2%

These error bounds may be considered typical of good practice and may be used when more specific data are not available.

CASE HISTORIES

The previous sections deal primarily with procedures for estimating the potential for triggering liquefaction. If triggering is likely to occur, then it becomes necessary to assess the consequences of liquefaction and what remedial measures to adopt. It is becoming more common in engineering practice to rely on deformation analysis as well as on the traditional factor of safety approach to evaluate the consequences of liquefaction and the options for cost-effective remedial measures.

These uses of deformation analysis will be demonstrated by two case histories from recent practice. The first case involves the evaluation of the potential for flow deformations in a tailings dam. The second case describes the analysis of the proposed nailing of the upstream slope of an embankment dam to firm foundation soils by driving piles across a thin weak horizontal layer in the upstream foundation.

ST. JOE STATE PARK TAILINGS DAM

Two tailings dams were constructed by the St. Joe Lead Company near Flat River, Missouri, from 1911 to 1965. These dams are designated Original Dam and Main Dam. When the mine closed, the entire processing facility was donated to the State of Missouri. The dams are situated within the zone of influence of the 1811-1812 New Madrid earthquakes. An assessment of the seismic stability of these dams was conducted in 1990 and reported by Vick et al. (1992). An idealized section of the Original Dam is shown in Fig. 8.

The liquefaction resistances of the materials in the dam against the design earthquake were obtained using the liquefaction resistance curve developed by Seed et al. (1985). The resistances so obtained were corrected for overburden pressure and for the initial static shear-stress ratio τ_{st}/σ'_{vo} using the correction factors K_σ and K_α given by Seed and Harder (1990).

The post-liquefaction flow deformations of the Original Dam (Fig. 8) were estimated using the program TARA-3FL (Finn and Yogendrakumar, 1989). The analysis is based on a Lagrangian formulation incorporating an adaptive mesh. The progress of deformations as the shear strength drops towards the residual value may be observed by comparing the deformed shapes of the dam in Figs. 9 and 10. Figure 9 shows the deformed shape when the undrained strength

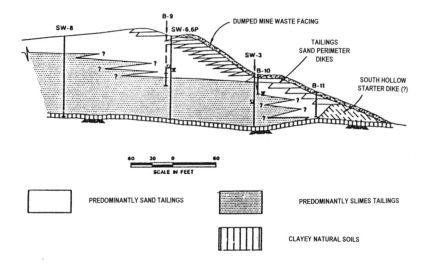

Fig. 8. Idealized Section of Original Dam

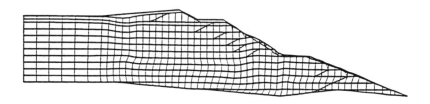

Fig. 9. Original Dam; Deformed Shape when Current Shear Strength of Slimes is 60% of Peak Strength.

in the slimes has dropped to 60% of the original peak value. Although substantial deformations have occurred the dam still maintains its integrity.

However, when the strength drops to 45% of the original undrained strength, massive deformations occur as shown in Fig. 10. This failure occurs long before the residual strengths are reached in all liquefiable elements.

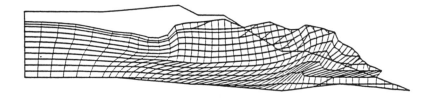

Fig. 10. Original Dam; Deformed Shape when Current Shear Strength of Slimes is 45% of Peak Strength.

ANALYSIS OF PILE REINFORCED SECTIONS OF SARDIS DAM

Previous studies have shown that the upstream section of Sardis Dam, Mississippi (Fig. 11), has the potential to undergo large deformations upstream along a layer of liquefiable clayey silt in the upper stratum clay should the design earthquake occur (Finn et al., 1991). The weak layer of clayey silt, 1.5 m (5 ft.) thick, is shown cross-hatched in Fig. 12. Due to liquefaction, the shear strength in this layer drops from an initial value in the region of 100 kPa (2000 psf) to a residual value of $S_{ur} = 0.075\ \sigma'_{vo}$ where σ'_{vo} is the effective overburden pressure. The method for restraining this deformation is to drive steel reinforced concrete piles through the upstream slope and the weak layer to an adequate penetration in the stronger layers below. In effect the upstream slope is nailed to the strong foundation layers (Fig. 12).

These piles must be designed to resist the combined effects of the pressures resulting from the restraint of the large deformations that might otherwise occur due to strength loss in the weak layer, and the moments and shears developed by seismic shaking during the design earthquake.

The key factors controlling the feasibility and cost of installing the pile to restrain the deformations are pile length and spacing, stiffness and strength of unliquefied soils surrounding the piles, residual strength of liquefied soils, the

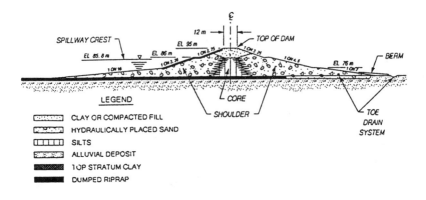

Fig. 11. Typical section of Sardis dam.

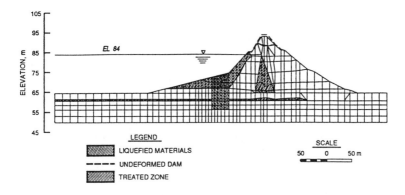

Fig. 12. Section of Sardis dam showing weak layer and restraining piles.

geometry of the structure, and the intensity of shaking before and after liquefaction has occurred. The ability to analyze such a complex problem while taking into account nonlinear behaviour of soil, potentially large strains in unremediated parts of the structure and a realistic interaction between piles and soil during both static and seismic loading is the essential requirement for determining the best location for the piles, an appropriate length and size and for categorizing the effects of soil properties.

Very little is known about the behaviour of piles under these complex conditions. Most of the evidence is from Japan where pile foundations have been severely damaged in liquefied ground as a result of ground displacements. However these piles were designed primarily for vertical static loading and both piles and the connections to the pile caps were inadequately designed to resist horizontal loading. Piles in Oakland Harbor were similarly damaged during the Loma Prieta Earthquake of 1989.

Detailed analytical studies were conducted to obtain data on the potential performance of piles resisting large upstream deformations in Sardis Dam and hence to provide a framework of understanding for the design of cost effective remediation using piles. These studies were conducted using the computer programs TARA-3 (Finn et al., 1986) and TARA-3FL (Finn and Yogendrakumar, 1989) which have proved useful in the analysis of flow deformations, settlements and the seismic response of earth structures and soil-structure interaction systems (Finn, 1988; Finn, 1994; Finn et al., 1994). The programs are not described here but detailed information is found in Finn (1988 and 1990).

The use of these analyses has allowed an acceptable displacement criterion to be used in defining satisfactory performance in addition to the common global factor of safety approach. This resulted in savings of several million dollars for the Sardis project. A problem with remediation measures designed solely on the basis of global factors of safety is that there is little understanding of what some of these factors mean in terms of displacements of the structure. Better and more cost-effective designs result from the implementation of Newmark's (1965) recommendation that the performance of earth structures be assessed using deformation criteria.

Analysis of Deformation Restraining Piles

A preliminary design option for restraining the flow deformations in Sardis Dam is to drive steel reinforced concrete piles, 0.6 m × 0.6 m (2 ft by 2 ft) in cross-section, through the upstream slope and the weak layer to a penetration of 4.6 m (15 ft) below the weak layer. The location for the piles shown in Fig. 12 was decided partly on analytical grounds and partly on practical considerations.

The piles are assumed to be at 3.7 m (12 ft) centres giving the plan layout shown in Fig. 13. Each row of piles normal to the axis of the dam is assumed to resist the potential movement of a plane strain section 3.8 m (12 ft) wide.

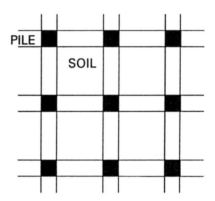

Fig. 13. Layout of restraining piles.

Analysis of pile response is conducted in two stages. First the effects of triggering residual strength in the layer of clayey silt are analyzed using the program TARA-3FL (Finn and Yogendrakumar, 1989). This analysis gives the moments and deflections in the piles arising from restraining the potential upstream movement of the embankment. Then the dam and piles, in the deformed condition, are subjected to earthquake shaking representative of design conditions using the program TARA-3 (Finn et al., 1986). This procedure for analysis simulates the condition where liquefaction and strength loss occur early in the earthquake. Subsequent studies confirm that this is the most conservative case.

Post-Liquefaction Deformation Analysis of Sardis Dam

The first step in the analysis is to construct the dam in layers to establish the initial stress conditions in the dam before remediation. Then the piles are inserted at the appropriate locations.

After liquefaction has occurred the initial stress state in the dam is no longer compatible with the stress-strain characteristics of the liquefied soils. The dam must therefore deform until a new equilibrium state is achieved. This involves very large strains in the unremediated section of Sardis Dam. To cope with these large strains, a Lagrangian formulation is used based on an adaptive mesh. When the upstream slope has been nailed to the foundation by piles, the inherent loss in resistance to lateral deformation along the weak layer of clayey silt is transformed to lateral load on the piles. The piles continue to deflect until an equilibrium state is achieved.

The distribution of lateral pressure on the lead pile after post-liquefaction equilibrium is reached is shown in Fig. 14.

The distribution of moments along the lead pile on the downstream side of the remediated zone is shown in Fig. 15. The peak moment is 1240 kNm (11,000 kip in), just above the weak layer.

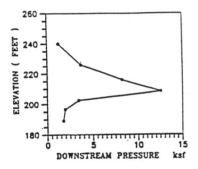
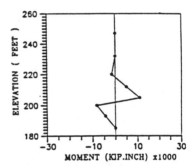

Fig. 14. Lateral pressures on lead piles after liquefaction (1 kip/ft^2 = 47.9 kPa; 1 ft = 0.3 m).

Fig. 15. Distribution of moments along lead piles after liquefaction (1 kip in = 113 Nm; 1 ft = 0.3 m).

The distributions of maximum shears and moments in piles along the remediated section are shown in Fig. 16 and Fig. 17, respectively. It is evident that the lead row of piles takes by far the largest moments and that these piles present the major challenge to effective design. The lead piles play such a dominant role because in the absence of a pile cap the only load transfer

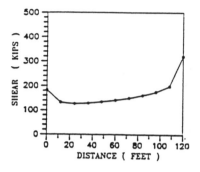

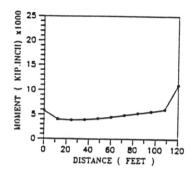

Fig. 16. Distribution of maximum
shears in piles along remediated
section (1 kip = 4.45 kN).

Fig. 17. Distribution of maximum
moments in piles along remediated
section (1 kip in = 113 Nm).

mechanism between the lead piles and the interior piles is the deflection of the
lead piles towards the interior piles and the transmission of the associated
increases in lateral pressures from one pile to another.

The deflection of piles in the leading row are shown in Fig. 18. The peak
deflection is of the order of 0.075 m (3 inches).

Seismic Analysis of Deformed Dam

The dynamic response of the remediated dam after triggering of residual
strength by liquefaction was evaluated using the program TARA-3 (Finn et al.,
1986). The input motions were the first 20 s of the S00E acceleration component
of the 1940 El Centro earthquake scaled to 0.2 g.

The time history of dynamic moments in the leading row of piles at the
top of the weak layer is shown in Fig. 19. The dynamic moments are plotted
starting from the post-liquefaction moment. This plot therefore, gives the total
moment at any instant during shaking. The peak moment occurring at t = 2.2 s
in the record is 1920 kNm (17,000 kip in) and the residual moment in the pile
after the earthquake is about 1020 kNm (9,000 kip in).

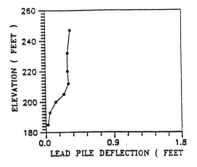

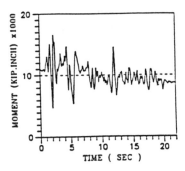

Fig. 18. Deflection pattern in the leading row of piles (1 ft = 0.3 m).

Fig. 19. Time history of total moment in the lead piles at top of weak layer (1 kip in = 113 Nm).

The time history of dynamic deformations at the top of the lead pile is shown in Fig. 20. The combined peak static and dynamic deflections of the lead pile upstream is of the order of 0.1 m (4 in).

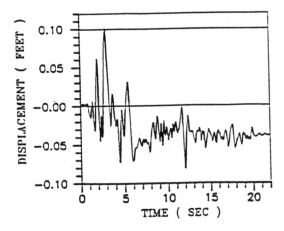

Fig. 20. Time history of deformations at the top of the lead piles (1 ft = 0.3 m).

The analyses demonstrated that potential large post-liquefaction deformations in Sardis Dam could be restrained. Both 2-D and 3-D parametric studies were conducted subsequently to optimize the location, design and spacing of the piles by nailing the upstream slope to the stable lower strata using piles. Pile nailing has now been adopted for stabilizing the upstream slope and construction is expected to start early in 1994.

CONCLUSIONS

Recent developments in the evaluation of liquefaction potential, the estimation of residual strength and the use of deformation analysis to assess the consequences of liquefaction and the evaluation of remedial measures have been reviewed. This paper should be viewed as an annotated guide to the referenced literature on issues that have a major impact on the cost of remediating structures with potential for liquefaction. The following inferences may be made from the current state of knowledge.

The correction factors K_σ and K_α used with the Seed liquefaction resistance chart can have a major impact on the potential for triggering liquefaction and the cost of remediation. Laboratory test data on reconstituted samples and the case history of Duncan Dam show clearly that these factors should be determined on a site specific basis. Reliance on average factors derived from undifferentiated global data can be grossly conservative in particular cases.

Residual strength, S_{ur}, is a function of effective overburden pressure σ'_{vo} In the case of Duncan Dam, $S_{ur} = 0.21\ \sigma'_{vo}$. At high overburden pressures, available data suggest that residual strength may be much higher than that suggested by the correlation between $(N_1)_{60}$ and S_{ur} proposed by Seed and Harder (1990).

The Duncan Dam study has indicated the feasibility and cost-effectiveness of retrieving frozen samples for direct laboratory evaluation of liquefaction resistance and residual strength in the case of fairly clean ands. In this case, data from the frozen samples precluded any need for remediation whereas current practice indicated the need for substantial remediation.

The time has arrived for a major re-assessment of procedures for evaluating the potential for triggering liquefaction, evaluating residual strength and conducting assessments of the consequences of liquefaction. Essential components of such an evaluation are fundamental laboratory research studies and more field studies like the Duncan Dam study which involve comparisons between the findings using current practice and findings based on tests on frozen samples recovered from various depths of overburden.

The use of deformation analysis gives a global representation of the post-liquefaction deformations. By adopting deformation criteria for the design of remedial measures, the designer has a much better opportunity to develop cost-effective remediation options in particular cases such as cases where extensive freeboard is available. In the Sardis Dam project, considerable savings were achieved by moving from a criterion based on a factor of safety to a criterion based on deformation.

The deformation approach is not advocated as a replacement for the safety factor approach but as a very useful complement. It allows a given factor of safety to be interpreted in terms of an associated level of deformation, and therefore can give guidance in the adoption of a more logical factor of safety for a particular job. The two approaches complement each other. In any specific situation based on the available data, confidence in the results of the deformation analysis and experience, the designer is free to give whatever weight he wishes to either of these two approaches.

The very detailed studies conducted on Sardis Dam showed that the pile nailing option would not have been feasible without the static and dynamic deformation analyses. Parametric studies using these techniques provided the data on which to base what proved to be the most cost-effective option by far for remediation of the dam.

ACKNOWLEDGEMENTS

Research on liquefaction and soil-structure interaction is supported by the National Science & Engineering Council of Canada under Grant #1498 to the lead author. Permission for publication of the material on Sardis Dam was granted by the Chief, U.S. Army Corps of Engineers.

REFERENCES

Byrne, P.M., A.S. Imrie and N.R. Morgenstern. (1993). "Results and Implications of Seismic Response Studies - Duncan Dam," Proceedings of the 46th Annual Canadian Geotechnical Conference, Saskatoon, Saskatchewan, Canada, pp. 271-281.

Finn, W.D. Liam (1994). "Seismic Safety Evaluation of Embankment Dams," Proc., Int. Workshop on Dam Safety Evaluation, Dam Engineering & ICOLD, Grindelwald, Switzerland, April 26-28, Vol. 4, pp. 91-135.

Finn W.D. Liam, R.H. Ledbetter and W.M. Marcuson III (1994). "Seismic Deformations in Embankments and Slopes," Proc., Symp. on Developments in Geotechnical Engineering - From Harvard to New Delhi, 1936-1994, Bangkok, Thailand, A.A. Balkema, Rotterdam.

Finn, W.D. Liam. (1990). "Seismic Response of Embankment Dams," Invited State-of-the-Art Paper, Dam Engineering, Vol. I, January, pp. 59-75.

Finn, W.D. Liam. (1988). "Dynamic Analysis in Geotechnical Engineering," Earthquake Engineering and Soil Dynamics II - Recent Advances in Ground Motion Evaluation, Geotech. Special Publication No. 20, ASCE, August, pp. 523-591.

Finn, W.D. Liam, R.H. Ledbetter, R.L. Fleming Jr., A.E. Templeton, T.W. Forrest and S.T. Stacy. (1991). "Dam on Liquefiable Foundation: Safety Assessment and Remediation," Proceedings, 17th International Congress on Large Dams, Vienna, pp. 531-553, June .

Finn, W.D. Liam and M. Yogendrakumar. (1989). "TARA-3FL - Program for Analysis of Liquefaction Induced Flow Deformations," Department of Civil Engineering, University of British Columbia, Vancouver, B.C., Canada.

Finn, W.D. Liam, M. Yogendrakumar, N. Yoshida and H. Yoshida. (1986). "TARA-3: A Program for Nonlinear Static and Dynamic Effective Stress Analysis," Soil Dynamics Group, University of British Columbia, Vancouver, B.C., Canada.

Harder, L.F. (1988). "Use of Penetration Tests to Determine the Cyclic Loading Resistance of Gravelly Soils During Earthquake Shaking," Ph.D. Thesis, Dept. of Civil Engineering, University of California, Berkeley.

Kuerbis, R.H. and Y.P. Vaid. (1989). "Undrained Behaviour of Clean and Silty Sand," Proceedings, 12th International Conference on Soil Mechanics and Foundation Engineering, Rio de Janeiro, Brazil, August.

Koester, J.P. (1992). "The Influence of Test Procedure on Correlation of Atterberg Limits with Liquefaction in Fine-Grained Soils," Geotechnical Testing Journal, GTJODJ, Vol. 15, No. 4, December, pp. 352-361.

Koester, J.P. (1990). Letter Report to Vicksburg District, U.S. Army Corps of Engineers.

Lo, R.C., E.J. Klohn and W.D. Liam Finn. (1991). "Shear Strength of Cohesionless Materials Under Seismic Loading," Proceedings of the 9th Pan-American Conference on Soil Mechanics and Foundation Engineering, Vina del Mar, Chile, Vol. 3, pp. 1047-1062.

McLeod, H., R.W. Chambers and M.P. Davies. (1991). "Seismic Design of Hydraulic Fill Tailings Structures," Proceedings of the 9th Pan-American Conference on Soil Mechanics and Foundation Engineering, Vina del Mar, Chile, Vol. 3, pp. 1063-1081.

Newmark, N.M. (1965) "Effects of Earthquakes on Dams and Embankments," 5th Rankine Lecture, Geotechnique, Vol. 15, No. 2, June, pp. 139-160.

Pillai, V.S. and R.A. Stewart. (1993). "Evaluation of Liquefaction Potential of Foundation Soils at Duncan Dams," Proceedings of the 46th Annual Canadian Geotechnical Conference, Saskatoon, Saskatchewan, Canada, pp. 237-258.

Poulos, S.J., G. Castro and J.W. France. (1985). "Liquefaction Evaluation Procedures," Journal of the Geotechnical Engineering Division, ASCE, Vol. 111, No. 6, June, pp. 772-792.

Seed, H.B., K. Tokimatsu, L.F. Harder and R.M. Chung. (1985). "Influence of SPT Procedures in Soil Liquefaction Resistance Evaluations," Journal of the Geotechnical Engineering Division, ASCE, Vol. 3, No. 12, December.

Seed, H.B. (1987). "Design Problems in Soil Liquefaction", Journal of the Geotechnical Engineering Division, ASCE, Vol. 3, No. 12, December.

Seed, R.B. and L.F. Harder Jr. (1990). "SPT-Based Analysis of Cyclic Pore Pressure Generation and Undrained Residual Strength," Proceedings, H. Bolton Seed Memorial Symposium, J.M. Duncan (Ed.), Vol. 2, May, pp. 351-376.

Troncoso, J.H. (1990). "Failure Risks of Abandoned Tailings Dams," Proceedings, International Symposium on Safety and Rehabilitation of Tailings Dams, ICOLD, Sydney, Australia, May, pp. 82-89.

Vaid, Y.P., E.K.F. Chung and R.H. Kuerbis. (1989). "Stress Path and Steady State," Soil Mechanics Series No. 128, Dept. of Civil Engineering, University of British Columbia, Vancouver, B.C., March.

Vaid, Y.P. and J. Thomas. (1994). "Post Liquefaction Behaviour of Sand," to appear in Proceedings, 13th International Conference on Soil Mechanics and Foundation Engineering, New Delhi, India, January.

Vick, S.G., R. Dorey, W.D. Liam Finn and R.C. Adams. (1992). "Seismic Stabilization of St. Joe Park Tailings Dams," submitted for review.

Wang, W. (1979). "Some Findings in Soil Liquefaction," Water Conservancy and Hydroelectric Power Scientific Research Institute, Beijing, China, August.

Earthquake-Induced Settlements in Silty Sands
for New England Seismicity

Cetin Soydemir[1]

Abstract

An attempt is made to extend the procedures by Tokimatsu and Seed (1987) and Ishihara and Yoshimine (1992) to predict seismically induced settlements in clean sands to silty sands by considering "appropriate corrections" in SPT blow-counts, $(N_1)_{60}$. A review of the basis of the "correction" concept and its implementation is presented as background. Results of the proposed procedure are compared with limited field data reported by others. Readily usable charts are developed for design practice to make approximate prediction of earthquake-induced settlements in silty sands for New England seismicity.

Introduction

Volumetric strains or settlements induced by ground shaking is one of the elements to be accounted for in earthquake-resistant design practice. Even "small" magnitudes of total and/or differential settlements may not be tolerated by certain special structures, and "moderate" settlements may not be acceptable or desirable for many types of common structures. Accordingly, prediction of seismically induced potential settlements in relatively loose, saturated "clean" sands received considerable attention during the past decade.

[1]Vice President, Haley & Aldrich, Inc., 58 Charles Street, Cambridge, MA 02141

Early works of Lee and Albaisa (1974) and Yoshimi et al. (1973) considered volumetric strains essentially resulting from reconsolidation or dissipation of excess pore pressures induced by cyclic loading. Thus, respective laboratory tests were typically carried only up to the stage of initial liquefaction (i.e., pore pressure ratio reaching to 100 percent) and the resulting volumetric strains were referred to as "initial liquefaction volumetric strains". Lee and Albaisa investigated the effect of mean grain size on initial liquefaction volumetric strain, and in a few tests they continued the tests for a number of cycles beyond initial liquefaction, observing that this resulted in substantially larger reconsolidation volumetric strains, referred to as "complete liquefaction volumetric strains". Shear strains induced by the cyclic loading were not accounted for in these tests. Pertinent data from Lee and Albaisa tests are summarized in Figure 1, indicating that volumetric strains would increase with increasing mean grain size at a given relative density.

A decade later, Tatsuoka et al. (1984) reported a significant observation - that the amount of volumetric strain in saturated sands subjected to cyclic loading would be significantly influenced by the magnitude of maximum shear strain induced during the post-initial liquefaction state. Tokimatsu and Seed (1987) taking into account the dominating effect of maximum shear strain, proposed a simplified procedure to predict volumetric strains for saturated clean sands, which has been widely adopted in current design practice. Ishihara and Yoshimine (1992) presented a similar simplified procedure which has generally confirmed the Tokimatsu and Seed method (Soydemir 1993). Selected data by Ishihara and Yoshimine are also incorporated in Figure 1 to demonstrate the significance of the maximum shear strain induced subsequent to initial liquefaction in producing volumetric strains.

Both Tokimatsu and Seed (1987) and Ishihara and Yoshimine (1992) methods are especially practical in design analyses since the saturated clean sands under study need only to be characterized by the respective Standard Penetration Test (SPT) blow-count $(N_1)_{60}$ profiles. It is again a practical necessity that attention is next be focused on sands having fines content (i.e., percent by weight passing No. 200 sieve) above about 5 percent, ranging from sands with little silt to silty sands. The paper attempts to develop a framework based on available methods and reported data for clean sands, which could serve to make approximate prediction of earthquake induced settlements in silty sands with a particular application to New England, a region of moderate seismicity.

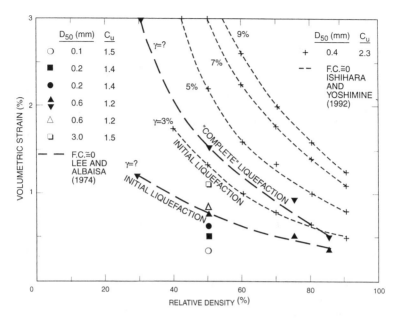

Figure 1. Data on Effect of Grain Size and Distribution on
Volumetric Strain in Clean Sands

Effect of Fines Content on Seismic Behavior

A silty sand is by definition a sand with fines content ranging from
about 5 percent to approaching as much as 50 percent. Thus, an
understanding of the clean sand behavior is a key element in extending
methodologies established for clean sands to silty sands in liquefaction
related problems.

Effect of fines content on liquefaction resistance of sands was
observed by Japanese engineers during the course of research undertaken
following the 1964 Niigata earthquake which caused extensive liquefaction
damage. Iwasaki et al. (1978) proposed a relationship between resistance
to liquefaction and SPT blow-counts which incorporated a "correction factor"
for fines content expressed as a function of the mean grain size, D_{50}.
Tatsuoka et al. (1980) based on laboratory test data, reported that silty
sands are considerably less vulnerable to liquefaction than clean sands with

similar SPT blow-counts. Tokimatsu and Yoshimi (1981) based on a comprehensive field study following the 1978 Miyagiken-oki earthquake, confirmed that this was indeed the case. Similarly, Zhou (1981) investigated the liquefaction/non-liquefaction sites after the 1976 Tangshan earthquake in China, and concluded that correlations between SPT blow-counts and liquefaction for clean sands would not be applicable for silty sands unless they are modified to allow for the fines content of the silty sands. Zhou suggested that for the same penetration resistance this allowance might be in the form of an increase in blow-counts whose magnitude would depend on the relative fines content.

Seed and Idriss (1982) and Seed et al. (1983) reviewed the data previously reported by the Japanese engineers and observed that the liquefaction envelope (i.e., cyclic stress ratio versus SPT blow-count corrected for effective overburden stress, N_1) for silty sands lies considerably above the liquefaction envelope for clean sands. They suggested that for a given cyclic stress ratio, the corresponding N_1 value for sands ($D_{50} >$ 0.25 mm) would be essentially equal to N_1 for silty sands ($D_{50} > 0.15$ mm) plus 7.5, that is a silty sand, as defined, would have the same resistance to initial liquefaction, even though its N_1 value is about 7.5 blow-counts smaller than the N_1 value of a clean sand. Tokimatsu and Yoshimi (1983, 1984) argued that using fines content as an index parameter relating to liquefaction resistance would be practically more advantageous than the mean grain size D_{50}, and proposed a range of "corrections" in the N_1 value as a function of fines content. Seed et al. (1985) based on a comprehensive review of field data available at the time, presented upgraded initial liquefaction envelopes for clean sands as well as sands with 15 and 35 percent fines content, which are at the present time utilized world-wide in design practice. Above referenced recommendations for fines content correction in $(N_1)_{60}$ values relative to resistance to initial liquefaction are summarized graphically in Figure 2.

Skempton (1986) considered the effect of fines content perhaps in a more fundamental way and noted that at approximately equal relative density, the SPT blow-count normalized for effective overburden stress, $(N_1)_{60}$, would be lower for fine sands than for coarse sands. Skempton's correlations between relative density and $(N_1)_{60}$ for coarse and fine sands were extended to silty sands by Seed et al. (1988), which projects lower $(N_1)_{60}$ values for silty sands than clean sands at a given relative density.

More recently, Seed (1986, 1987) proposed the use of $(N_1)_{60}$ value to determine the residual undrained (steady state) strength in conducting post-liquefaction stability evaluation of embankment dams, and provided "correction" values for $(N_1)_{60}$ as a function of fines content to obtain equivalent clean sand $(N_1)_{60}$ values for silty sands. These "correction" values are also incorporated in Figure 2.

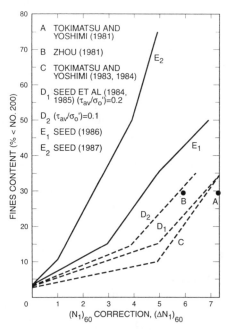

Figure 2. $(N_1)_{60}$ Correction   Values   for Respective Fines Content

The concept of utilizing $(N_1)_{60}$ values in evaluation of seismically induced phenomena such as resistance to initial liquefaction, volumetric strains and residual undrained strength for clean sands is a "powerful" tool in design practice since it bypasses extremely difficult sampling and laboratory testing procedures. Equally, to extend established relationships for clean sands to silty sands by "appropriate correction" of $(N_1)_{60}$ values (i.e., the equivalent clean sand concept), provides an "enormous" convenience to the design engineer. However, it has been the author's experience that the proposed "family of correction" values, also designated as $\Delta(N_1)_{60}$, are not clearly understood in general design practice and at times lead to misinterpretation (e.g., Ferritto 1992). Seed and Harder (1990) provided a timely clarification on this matter indicating that $(N_1)_{60}$ correction values for the "triggering" (i.e., initial liquefaction) and "post-triggering" (i.e., residual undrained strength) analyses are inherently different as they relate to different phenomena, and the appropriate "correction" values for fines content should be used for the particular problem under consideration.

It is in the context of the review presented above that the potential effect of fines content be examined in developing volumetric strain versus $(N_1)_{60}$ relationships. It may be suggested that volumetric strains are affected by mechanisms at work both during the "triggering" phase (i.e., pore pressure build-up) and the "post-triggering" phase (i.e., maximum shear strain). It may also be argued that progressive build-up of pore pressures as well as shear strains occur during the period of ground shaking, and thus the "triggering" phase. At the present there are no systematic laboratory testing data available on silty sands as are reported on clean sands (e.g., Ishihara and Yoshimine 1992). Equally, there are no systematic field data accumulated on seismically induced volumetric strains relating to $(N_1)_{60}$ for silty sands as on clean sands (e.g., Tokimatsu and Seed 1987; Ishihara and Yoshimine 1992). In this study, it has been adopted that seismically induced volumetric strains are basically a "triggering" phase occurrence.

Design Earthquake for New England

New England is an intra-plate region of moderate seismicity. Major historical earthquakes in New England and suggested source mechanisms were reviewed by Soydemir (1987). Based on the recommendations of the Massachusetts Seismic Design Advisory Committee (1983), a M=6.5 event with a maximum horizontal ground acceleration of 0.12 g was adopted as the design earthquake in developing the earthquake-resistant design provisions of the current Massachusetts State Building Code (1991) which also includes criteria for initial liquefaction and seismically induced earth pressures on retaining walls. It has been previously suggested that a M=6.5, a_{max} = 0.12 g earthquake would be representative of the New England region in consideration of geotechnical-earthquake engineering problems (Soydemir 1987).

Settlements for New England Seismicity

Prediction of seismically induced settlements in glaciofluvial, lacustrine and more recent floodplain deposits of saturated clean sands as well as silty sands is a common design problem in New England. This includes both diagnostic analyses and developing remedial measures, if and as necessary (Soydemir et al. 1994). Based on the procedures proposed by Tokimatsu and Seed (1987) and Ishihara and Yoshimine (1992), Soydemir (1993) developed readily usable correlations (i.e., volumetric strains versus blow-count and depth) to predict seismically induced settlements in clean sands for New England. The need to extend such correlations to include silty sands has been the primary motivation for this study.

At initial liquefaction (i.e., pore pressure ratio of 100 percent) Ishihara and Yoshimine (1992) established that a maximum shear strain of 2 to 4 percent would occur in clean sands resulting in volumetric strains ranging between about 0.5 to 1.5 percent over the practical range of in-situ relative densities (Figure 1). This is in general agreement with the data reported by Lee and Albaisa (1974). It is proposed that volumetric strain at initial liquefaction may be used as a "benchmark" in settlement analyses. Based on the data reported by Lee and Albaisa (1974), it may further be suggested that the "benchmark" volumetric strain for a silty sand would be smaller than the "benchmark" volumetric strain for a clean sand at a particular relative density.

Subsequent to initial liquefaction stage with continuing ground shaking, progressively larger shear strains would be induced leading to larger volumetric strains. Ishihara and Yoshimine (1992) correlated magnitude of maximum shear strain to a "factor of safety" parameter defined as the ratio of the limiting cyclic stress ratio at initial liquefaction to the average cyclic stress ratio induced by ground shaking. Thus "factor of safety" could assume values below 1.0 as well as above 1.0. Ishihara and Yoshimine established that at a "factor of safety" of about 2.0 and above, insignificant volumetric strain would occur. Following this procedure, Figure 3 was developed earlier by Soydemir (1993) for clean sands.

It is suggested on an approximate basis that the Ishihara and Yoshimine (1992) "factor of safety" versus $(N_1)_{60}$ correlations proposed for clean sands may be adopted for silty sands, which has been followed in this study, to project volumetric strains above the "benchmark" volumetric strains. Accordingly, Figures 4 and 5 were developed corresponding to fines

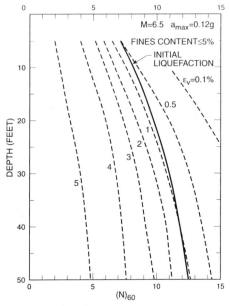

Figure 3. Volumetric Strain vs $(N)_{60}$ at a Depth
for Clean Sands

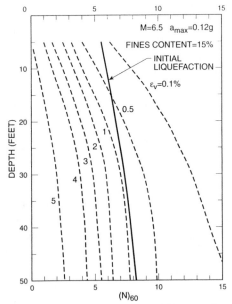

Figure 4. Volumetric Strain vs $(N)_{60}$ at a Depth
for Fines Content of 15 Percent

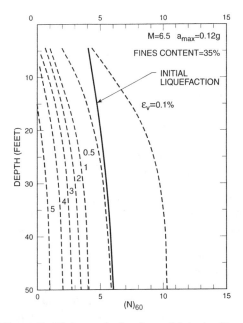

Figure 5. Volumetric Strain vs (N_{60}) at a Depth for Fines Content of 35 Percent

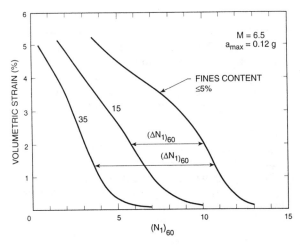

Figure 6. Average Volumetric Strain vs ($N_1)_{60}$ Relationships for Different Fines Content

content of 15 and 35 percent, respectively. In Figures 3, 4 and 5, in addition to the volumetric strain contours, envelopes corresponding to initial liquefaction are incorporated. These envelopes were determined based on the procedure and correlations provided by Seed et al. (1985).

It should be noted that Figures 3, 4 and 5 would be utilized by directly entering the figures with the actual field SPT blow-count data if a safety hammer, $(N)_{60}$, has been utilized in the test borings. However, the donut hammer is still widely used in New England, and if that is the case, the field data, $(N)_{45}$, should be multiplied by 0.75 to obtain the equivalent$(N)_{60}$ value (Seed et al. 1985). This is to account for the different efficiencies of the two particular types of hammers.

Figure 6 was deduced from Figures 3, 4 and 5 to establish the effect of fines content on volumetric strain. The figure also establishes "correction" values for $(N_1)_{60}$, if an equivalent clean sand approach is followed to predict volumetric strains for silty sands with fines content up to about 35 percent. It may be observed that the "correction" values besides fines content are also dependent on the particular $(N_1)_{60}$ value which is being corrected. For the broad "middle range" of $(N_1)_{60}$ values of practical interest, the "correction" values, $\Delta(N_1)_{60}$, are in close range of the correction values proposed by Seed et al. (1985) in establishing resistance to initial liquefaction for silty sands (Figure 2).

Field Observations

O'Rourke et al. (1991) reported the results of a study comparing the observed and predicted seismically induced settlements at the Marina District, San Francisco, following the 1989 Loma Prieta earthquake. They have established close agreements in those areas underlain by clean sands (fills and natural deposits) based on the Tokimatsu and Seed (1987) procedure; however, the same procedure overestimated the settlements nearly by 100 percent for the hydraulic-fill sands with about 15 percent fines content.

O'Rourke et al. (1991) suggested that a "correction" of about 5 blows be applied for the hydraulic-fill sands to bring up the average field value of $(N_1)_{60} = 4.5$ to $(N_1)_{60} = 9.5$ based on the data reported by Seed et al. (1991) and Lee and Albaisa (1974), which then yielded a settlement magnitude consistent with the measured settlements (i.e., a volumetric strain of

2.7 percent). It may be noted in Figure 6 that for $(N_1)_{60}$ = 4.5, the corresponding volumetric strain for clean sand is about twice that of sand with 15 percent fines content. Similarly, Figure 6 indicates that a "correction", $\Delta(N_1)_{60}$, of about 5 be made for sand with 15 percent fines content to establish an equivalent clean sand $(N_1)_{60}$ value. There is also general agreement in actual magnitudes of volumetric strains reported by O'Rourke et al. and those predicted in Figure 6, where further slight adjustments for the earthquake magnitude and acceleration are to be made for a more rigorous comparison.

Similarly, Egan and Wang (1991) reported measured and predicted settlements at Treasure Island, San Francisco, also after the 1989 Loma Prieta earthquake. Settlements calculated for the hydraulic-fill sands by the Tokimatsu and Seed (1987) procedure were again found to be overestimated by a factor of about two. Following a "modified" approach which took into account the influence of grain size based on the Lee and Albaisa (1974) data Egan and Wang reported that they obtained a reasonable agreement with the measured settlements. No specific data was provided on the fines content of the hydraulic-fill sand for a comparative evaluation.

Conclusions

Prediction of seismically induced settlements in clean sands can be undertaken at the present following the procedures proposed by Tokimatsu and Seed (1987) and Ishihara and Yoshimine (1992). However, the design engineer often has to deal with silty sands in diagnostic settlement analyses as well as in developing criteria for remedial measures for improvement of sites underlain by relatively loose silty sand deposits and/or fills. The study attempts to provide readily usable charts for the design engineer to make approximate predictions of earthquake-induced settlements in silty sands with particular application to New England. However, the framework presented may be implemented to regions characterized by a different "design earthquake". The approach followed is based on the concept of converting $(N_1)_{60}$ field data for silty sands to equivalent $(N_1)_{60}$ values for clean sands by applying a "correction" on the field $(N_1)_{60}$ data. "Correction" for the fines content concept initially introduced by the Japanese engineers and subsequently formulated by Seed et al. (1985) and Seed (1987) in problems of prediction of initial liquefaction and residual (steady state) shear strength has been "extended" for the prediction of volumetric strains in silty sands.

It has been the author's experience that at the present there is some inconsistency and at times misuse in design practice regarding the application of "correction" values, $\Delta(N_1)_{60}$, to different liquefaction problems. The critical element in this application is to establish beforehand whether the particular problem (e.g., initial liquefaction, residual strength, volumetric strain) belongs to the "triggering" or the "post-triggering" phase (Seed and Harder 1990). It is amplified that the volumetric-strain problem is a "triggering" phase problem.

Acknowledgements

Prof. R. Dobry of Rensselaer Polytechnic Institute and Prof. M. Evans of Northeastern University have provided insight during the course of the study. Ms. A. Welch and Ms. C. Catwell, both of Haley & Aldrich, prepared the figures and the manuscript, respectively. These contributions are greatly appreciated.

References

Egan, J.A and Wang, Z.L. (1991). "Liquefaction related ground deformation and effects on facilities at Treasure Island, San Francisco, during the 17 October 1984 Loma Prieta earthquake." Proc. Third Japan-U.S. Workshop on Earthquake Resistant Design of Lifeline Facilities and Countermeasures for Soil Liquefaction., Tech. Report NCEER-91-0001, Feb., 57-76.

Ferrito, J.M. (1992). "Ground motion amplification and seismic liquefaction: A study of Treasure Island and the Loma Prieta earthquake." Naval Civil Engrg. Lab,. Tech Note N-1844, Prot Hueneme, CA.

Ishihara, K. and Yoshimine, M. (1992). "Evaluation of settlements in sand deposits following liquefaction during earthquakes." Soils and Found., JSSMFE, 32(1), 173-188.

Iwasaki, T., Tatsuoka, F., Tokida, K., and Yasuda, S. (1978). "A practical method for assessing soil liquefaction potential based on case studies at various sites in Japan." Proc. Second Int. Conf. Microzonation, San Francisco, CA, 2, 885-896.

Lee, K.L. and Albaisa, A. (1974). "Earthquake-induced settlements in saturated sands." J. Soil Mech. Found., Div., ASCE, 100(4), 387-400.

Massachusetts Seismic Design Advisory Committee (1983). "Proposed changes to earthquake design sections." J. Boston Soc. of Civil Engrg./ASCE, 69(2), 209-234.

Massachusetts State Building Code (1991), Fifth Edition, 780 CMR, Boston, MA.

O'Rourke, T.D., Gowdy, T.E., Stewart, H.E., and Pease, J.W. (1991). "Lifeline performance and ground deformation in the Marina during 1989 Loma Prieta earthquake." Proc. Third Japan-U.S. Workshop on Earthquake Resistant Design of Lifeline Facilities and Countermeasures for Soil Liquefaction. Tech. Report NCEER-91-0001, Feb., 129-146.

Seed, H.B. (1986). "Design problems in soil liquefaction." Report No. UCB/EERC-86/02, Univ. of Calif., Berkeley, CA.

Seed, H.B. (1987). "Design problems in soil liquefaction." J. Geotech. Engrg. Div., ASCE, 113(8), 861-878.

Seed, H.B. and Idriss, I.M. (1982). Ground Motions and Soil Liquefaction during Earthquakes. Monograph Series, Earthquake Engrg. Res. Ins., Berkeley, CA.

Seed, H.B., Idriss, I.M. and Arango, I. (1983). "Evaluation of liquefaction potential using field performance data." J. Geotech. Engrg. Div., ASCE, 109(3), 458-482.

Seed, H.B., Tokimatsu, K., Harder, L.F., and Chung, R. (1985). "The influence of SPT procedures in soil liquefaction resistance evaluations." J. Geotech Engrg. Div., ASCE, 111(12), 1425-1445.

Seed, H.B., Seed, R.B., Harder, L.F., and Jong, H-L. (1988). "Re-evaluation of the slide in the Lower San Fernando Dam in the earthquake of Feb. 9, 1971." Report No. UCB/EERC-88/04, Univ. of Calif., Berkeley, CA.

Seed, R.B. and Harder, L.F. (1990). "SPT-based analysis of cyclic pore pressure generation and undrained residual strength." Proc. H.B. Seed Memorial Symposium, Berkeley, CA, (2), 351-376.

Skempton, A.W. (1986). "Standard penetration test procedures and effects in sands of overburden pressure, relative density particle size, ageing and overconsolidation." Geotechnique, 36(3), 425-447.

Soydemir, C. (1987). "Liquefaction criteria for New England considering local SPT practice and fines content." Proc. Fifth Canadian Conf. on Earthq. Engrg., Ottawa, Canada, 519-525.

Soydemir, C. (1993). "Earthquake induced settlements for New England seismicity." Proc. National Earthq. Conf., Memphis, TN.

Soydemir, C. Stulgis, R.P., and F. Swekosky. (1994). "Liquefaction/vibro-compaction remedial design criteria for New England seismicity." Paper submitted to Fifth U.S. Earthq. Engrg. Conf., Chicago, IL.

Tatsuoka, F., Iwasaki, T., Tokida, K., Yasuda, S., Hirose, M., Imai, T., and Kon-no, M. (1980). "Standard penetration tests and soil liquefaction potential evaluation." Soils and Found., JSSMFE, 20(4).

Tatsouka, F., Sasaki, T., and Yamada, S. (1984). "Settlement in saturated sand induced by cyclic undrained simple shear." Proc. Eight World Conf. Earthq. Engrg., San Francisco, CA, 95-102.

Tokimatsu, K. and Yoshimi, Y. (1981). "Field correlation of soil liquefaction with SPT and grain size." Proc. Int. Conf. Recent Advances in Geot. Earthq. Eng. and Soil Dynamics, St. Louis, MO, 1, 203-208.

Tokimatsu, K. and Yoshimi, Y. (1983). "Empirical correlation of soil liquefaction based on SPT N-value and fines content." Soils and Found., JSSMFE, 23(4), 56-74.

Tokimatsu, K. and Yoshimi, Y. (1984). "Criteria of soil liquefaction with SPT and fines content." Proc. Eight World Conf. Earthq. Engrg., San Francisco, CA, 255-262.

Tokimatsu, K. and Seed, H.B. (1987). "Evaluation of settlements in sands due to earthquake shaking." J. Geotech. Engrg. Div., ASCE, 113(8), 861-878.

Yoshimi, Y., Kuwabara, F. and Tokimatsu, K. (1973). "One-dimensional volume change characteristics of sands under very low confining stresses." Soils and Found., JSSMFE, 15(3), 51-60.

Zhou, S.G. (1981). "Influence of fines on evaluating liquefaction of sand by CPT." Proc. Int. Conf. Recent Advances in Geot. Earthq. Eng. and Soil Dynamics, St. Louis, MO, 1, 167-172.

Liquefaction of Artificially Filled Silty Sands

Susumu Yasuda[1]
Kazue Wakamatsu[2]
Hideo Nagase[1]

Abstract

Liquefaction strength of artificially filled silty sands was studies by conducting undrained cyclic triaxial tests and case studies on liquefaction histories of reclaimed lands. The undrained cyclic triaxial tests showed that liquefaction strength of a sand containing many fine particles increases faster than the liquefaction strength of a fine sand. The case studies demonstrated also that aging effects on liquefaction strength becomes strong with contents of fine particles.

Introduction

During the 1987 Chibakentoho-oki earthquake, liquefaction was induced in many reclaimed lands along Tokyo Bay. The grain size of the boiled soils in the reclaimed lands was quite a bit small. In Japan, it has been considered that such fine soil, namely silty sand or silt, is hard to liquefy due to cohesion.

Then, several studies on liquefaction strength of artificially filled silty sands have been conducted by undrained cyclic triaxial tests on undisturbed and reconstituted samples, and case studies on liquefaction histories and SPT N-values of artificially reclaimed grounds.

Liquefaction of Silty Sands During the 1987 Chibakentoho-oki Earthquake

The Chibakentoho-oki earthquake, with a magnitude of 6.7, occurred on December 17, 1987, almost 70 km southeast of Tokyo and produced liquefaction at many sites as shown in Fig.1. According to the authors' site

---

1 Associate Professor, Kyushu Institute of Technology, Tobata, Kitakyushu, 804 JAPAN
2 Guest Researcher of Advanced Research Center for Science and Engineering, Waseda University, Kikuicho, Shinjuku, Tokyo, 162 JAPAN

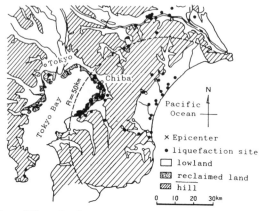

Fig.1 Sites of liquefaction caused by the 1987 Chibakentoho-oki
earthquake

investigation, the liquefaction sites were classified into four groups (Yasuda et al., 1989), (1) reclaimed land along Tokyo Bay, (2) lowland along the Pacific Ocean, (3) alluvial lowland along the Tone River, and (4) filled land in valleys.   Among these groups, the first group, reclaimed land along Tokyo Bay, was very interesting, because very fine silty sands were liquefied.

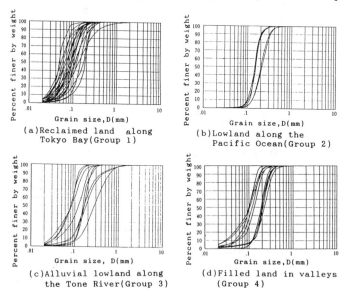

Fig.2 Grain size distribution curves of soils boiled by the
1987 Chibakentoho-oki earthquake

Figure 2 compares the grain size distribution curves of boiled soils obtained from the four groups of liquefaction sites mentioned before. The grain size of the boiled soils in the reclaimed land is quite a bit smaller than those of the boiled sands in other groups. In Japan, it has been considered that such fine soil, namely silty sand or silt, is hard to liquefy due to cohesion. In fact, undrained cyclic strength of undisturbed alluvial silt is fairly strong.

Liquefaction Strength of Reconstituded Samples

To study the reason why such very fine silty sand had liquefied, undrained cyclic triaxial tests were performed on almost 30 reconstituded samples (Yasuda et al., 1993). The tested samples, described in Table 1, are classified into the following seven groups:

Table 1 Tested samples

No.	Sample	D50(mm)	FC(%)	PC(%)	Ip	Dr(%)	R1	Roundness
1	Boiled sand , Anegasaki	0.084	42.0	16.0	NP	29.1	0.172	0.335
2	ditto , Sodegaura	0.120	30.8	15.0	NP	34.3	0.166	0.310
3	Volcanic ash sand , Kagoshima , A	0.212	19.8	8.0	NP	55.7	0.122	0.175
4	Hill sand , Kisarazu	0.130	16.5	10.5	7.7	80.5	0.210	0.390
5	ditto , Sengenyama	0.220	4.0	3.0	NP	83.2	0.170	0.275
6	Decomposed Granite , Kokura	0.150	39.1	14.0	12.3	93.3	0.210	0.290
7	ditto , Kaho	1.650	8.3	4.0	NP	86.9	0.189	0.335
8	Wast soil , Tokyo , A	0.210	21.7	12.2	NP	66.7	0.190	
9	ditto , B	0.170	34.6	20.0	4.1	63.0	0.237	
10	ditto , C	0.138	39.0	27.0	8.4	75.7	0.242	
11	ditto , D	0.240	27.2	17.8	8.3	59.5	0.268	
12	Dredged sand , Kanda	0.037	67.8	28.0	15.2	93.2	0.259	0.385
13	ditto , Fukuoka	1.000	19.0	10.0	NP	90.0	0.178	0.305
14	Toyoura sand , Yamaguchi	0.208	0.0	0.0	NP	35.3	0.151	0.395
15	Decomposed Granite , Yamaguchi , O	0.440	32.0	13.2	12.4	105.2	0.200	0.365
16	ditto , A	2.400	5.5	4.0	−	85.0	0.148	0.375
17	ditto , B	0.180	32.0	12.1	−	130.7	0.140	0.355
18	ditto , C	0.024	69.8	22.5	−	111.0	0.165	
19	ditto , F	0.009	100.0	34.0	−	129.7	0.195	
20	Boiled sand , MARINA	0.162	9.0	0.0	NP	42.2	0.150	0.450
21	ditto , RICHMOND	0.053	75.2	6.0	NP	42.8	0.165	0.380
22	ditto , ARAMEDA,RL	0.162	15.0	8.0	NP	58.8	0.136	0.485
23	ditto , ARAMEDA,AP	0.067	54.8	7.0	NP	56.5	0.162	0.385
24	ditto , PAHARO RIV	0.135	28.1	7.0	NP	40.7	0.185	0.395
25	ditto , MOSS LD,MARINA	0.110	6.8	4.0	NP	45.3	0.167	0.435
26	ditto , MOSS LD,PARK	0.230	29.3	7.0	NP	40.6	0.148	0.380
27	Volcanic ash sand , Kagoshima , B	0.232	24.0	11.0	NP	64.4	0.108	0.220

(1) boiled sands in artificially filled lands resulting from the 1987 Chibakentoho-oki earthquake and the 1989 Loma Prieta earthquake --- Samples No.1, 2, and 20 to 26, respectively.
(2) hill sands which have been used for artificial fill material in Tokyo Bay --- Samples No.4 and 5
(3) decomposed granite soils, named "Masado", which are used for artificial fill material in western Japan --- Samples No.6, 7, and 15 to 19
(4) dredged sands, which are most commonly used for artificial fill material in Japan --- Samples No.12 and 13 (Fukuoka Prefecture)
(5) volcanic ash sands, named "Shirasu", which are used for artificial fill material in southern Japan --- Samples No.3 and 27
(6) waste soils produced during the construction of structures --- Samples No.8 to 11
(7) Toyoura standard sand, which is a clean fine sand --- Sample No.14

For Samples No.16 to 19, a special technique for sample preparation was applied to disperse the grains according to coarseness. A water channel, was filled with water to a depth of 15 cm. Then, a mixture of the original sample, No.15, and water was poured into the channel from the left end of the channel. Coarse grains were deposited close to the left end and fine grains flowed away and were deposited far from the left end. Samples from three parts: A, B and C were taken and used for the undrained cyclic triaxial tests. Sample F was composed of fine particles, of less than 74 μ in diameter, prepared by passing the original soil through a sieve with a mesh of 74 μ.

There are two methods to reconstitute a sample. One is to make the sample with a constant relative density, and the other procedure is to reconstitute it with a constant energy compacting method. In this study, the latter method was adopted because the method of filling reclaimed lands seemed to be similar to this one. The soils were dried first, then poured into a mold from a nozzle with a constant drop height, as shown in Fig.3. The height was adjusted to 30 cm to duplicate in-situ density. After saturation and consolidation, a cyclic load was applied to the specimen under an effective confining pressure of 49 kPa.

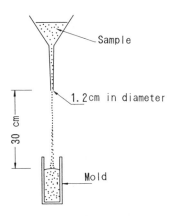

Fig.3 Diagram of apparatus used to prepare specimens

Test results of all samples are summarized in Table 1. Among these results, the relationships between liquefaction strength, R_ℓ ($N_\ell=20$, DA=5%), and fine content of less than 74 μ, Fc, are shown in Fig.4. The liquefaction strength increases if fine content increases. However, the rate of increase is not large. Moreover, scatter is fairly large. Fig.5 and Fig.6 show the relationship between liquefaction strength and clay content of less than 5 μ, Pc, and the relationship between liquefaction strength and Plasticity Index, Ip, respectively. Scatters in these figures are smaller than that in Fig.4. As shown in these figures, if a sand contains fairly many fine particles, the liquefaction strength of artificially filled deposit composed of such sand is not so large. This may be due to the short perild of time which had elapsed since filling. Namely, the structure of the grains of the reclaimed soil may have still been unstable because not enough time had elapsed to make stabilization or cementation sufficient to resist liquefaction.

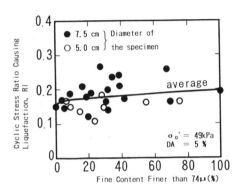

Fig.4 Relationship between liquefaction strength and fine content

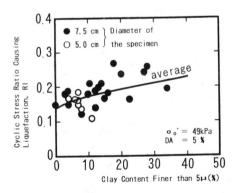

Fig.5 Relationship between liquefaction strength and clay content

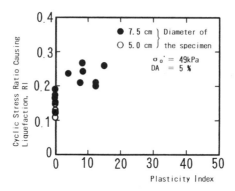

Fig.6 Relationship between liquefaction strength and plasticity index

Liquefaction Strength of Reclaimed Silty Sands and Alluvial Silty Sands

As mentioned above, the liquefaction strength of artificially filled silty sands is not so large. However, according to many experiences, the liquefaction strength of a natural deposit of silty sands is fairly large. This difference seems to be due to aging. For the effect of aging on liquefaction strength, Troncoso, Ishihara and Verdugo (1988) conducted cyclic triaxial tests on several tailing materials with different ages and clarified that the liquefaction strength increases with ages. Kimura, Tatsuoka and Pradhan (1986) also studied the aging effect of reconstitute samples.

To clarify the aging effect of silty sands, the liquefaction strength of ten undisturbed samples taken from alluvial sandy layers and reclaimed sandy layers were compared. The undisturbed samples were taken at five sites in Fukuoka City as shown in Fig.7. There are three kinds of loose sandy soils in Fukuoka City : (1) well graded alluvial silty sand, (2) poor graded dune sand, and (3) well graded reclaimed silty sand. The grain size distribution curves were compared in Fig.8. The liquefaction strengths, R_ℓ, on these samples obtained by the undrained cyclic triaxial tests were plotted with fine content finer than 74 μ and clay content finer than 5 μ in Fig.9 and Fig.10, respectively. As shown in these figures, the liquefaction strength of alluvial sandy soil was greater than that of reclaimed sandy soil in the same fine content or clay content. The point "Q" was the data of the reconstituted sample of a reclaimed sample. And the shaded zones in Fig.9 and Fig.10 show the spheres of the data on the reconstituted samples shown in Fig.4 and Fig.5. The point "Q" fell in the shaded zone. Therefore, it can be said that the liquefaction strength of reconstituted sample is lower than that of reclaimed sandy soil under the same fine or clay content.

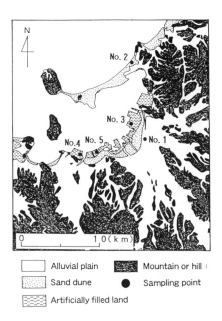

Fig.7 Geological map and sampling sites in Fukuoka City in Japan

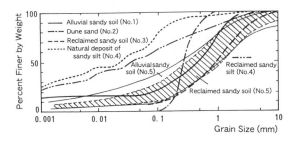

Fig.8 Grain size distribution curves of undisturbed samples

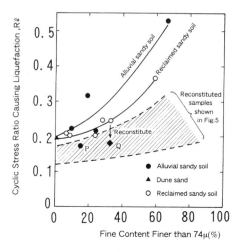

Fig.9 Relationship between liquefaction strength and fine content (Sample "P" was taken just from the boundary between the alluvial and reclaimed layers)

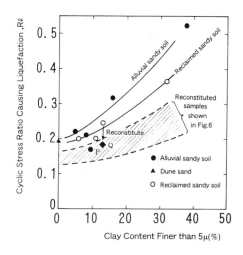

Fig.10 Relationship between liquefaction strength and clay content (Sample "P" was taken just from the boundary between the alluvial and reclaimed layers)

Increase of Liquefaction Strength of Artificially Filled Silty Sands with Time

To make clear the increase of liquefaction strength of artificially filled silty sands, case studies on liquefaction histories and SPT N-values of artificially reclaimed grounds were carried out.

In Japan many artificially reclaimed lands have been constructed along coasts from 19th century. As reclaimed soils are sensitive to liquefaction, liquefaction has been induced in many reclaimed lands during past earthquakes. Figure 11 shows the liquefied reclaimed lands during past earthquakes from 1855 to 1987. The number of the liquefied reclaimed lands reaches about 200.

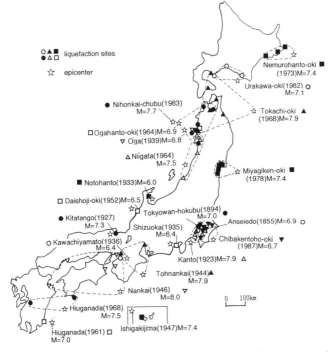

Fig.11 Locations of reclaimed lands liquefied during earthquakes from 1855 to 1987

Then, by studying the constructed years of the reclaimed lands with historical literatures, the terms from the years of construction to the years of occurrence of liquefaction, Ta, were counted. And the peak surface accelerations at those lands during the earthquakes which induced liquefaction, Ac, were evaluated using appropriate attenuation equations. At most of the reclaimed lands have the experiences of non-liquefaction during other earthquakes, the peak surface accelerations at the reclaimed lands during the other earthquakes were also evaluated, and the terms from the construction to the earthquakes were also counted.

Figure 12 shows relationships between the terms from the construction of the reclaimed lands to the earthquakes, Ta, and the peak accelerations at the reclaimed lands induced during the earthquakes, Ac. As shown in this figure, not only the data in Japan but also the data in the reclaimed lands along San Francisco Bay during the 1906 San Francisco earthquake and the 1989 Loma Prieta earthquake, and the datum in Lazaro Cardenas in Mexico during the 1985 Mexico earthquake, were also included. The liquefied events and non-liquefied events can be roughly separated with a boundary shown in broken line. Therefore, it can be said the minimum peak surface acceleration which induce liquefaction increases with time after reclamation. This means liquefaction strength of reclaimed soil increases with time due to aging effect.

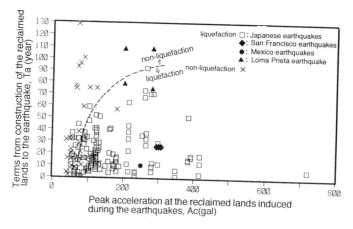

Fig.12 Relationships between Ta and Ac for all reclaimed lands

To make clear the influence of fine contents to the aging effects of reclaimed soils, the data of several reclaimed lands which were clear to be filled with clean sands or silty sands, were extracted from Fig.12. These selected reclaimed lands with clean sands were located at Aomori, Noshiro, Funakawa, Akita, Niigata and San Francisco which are filled with dune sands. On the contrary, reclaimed lands along Tokyo Bay were selected as the lands with silty sands. Figure 13 and Fig.14 show the relationships between the terms from construction of the reclaimed lands to the occurrence of the earthquakes, Ta, and peak accelerations during the earthquakes, Ac. By comparing two figures, the minimum peak surface acceleration which induce liquefaction was almost same between reclaimed lands with fine sands and those with silty sands if the term after reclamation is short. However, the minimum peak surface acceleration which cause liquefaction at the reclaimed lands with silty sands increases with time faster than that at the reclaimed lands with clean sands. This means that the aging effect on liquefaction strength becomes strong with contents of fine particles.

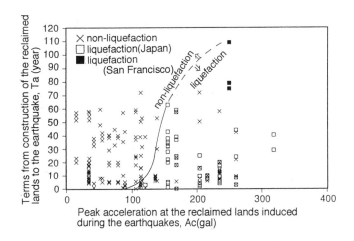

Fig.13 Relationships between Ta and Ac for reclaimed lands of clean sands

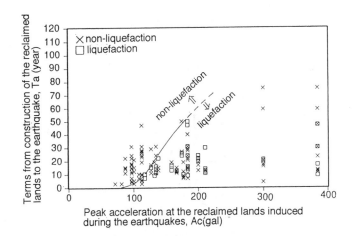

Fig.14 Relationships between Ta and Ac for reclaimed lands of silty sands

Increase of SPT N-values of Artificially Filled Silty Sands with Time

Along Tokyo Bay, many artificially reclaimed lands have been constructed with dredged soils from the bottom of the bay or with cut soil from sorrounding hills. Ages of reclamation of these lands are clearly recorded and a lot of soil investigations, by mainly Standard Penetration Tests, have been carried out. Then, relationships between SPT N-values and the terms from the construction of reclaimed lands to the year when the Standard Penetration Tests had been carried out, Tn, were studied. The measured SPT N-values were normalized in an effective overburden pressure of 0.6kgf/cm² (58.8 kPa) by utilizing the Meyerhof's formula, as:

$$\frac{N_n}{0.7+(\sigma_v')_n} = \frac{N_m}{0.7+(\sigma_v')_m} \qquad (1)$$

where, N_m: measured SPT N-value

$(\sigma_v')_m$: effective overburden pressure at the tested depth

N_n: normalized SPT N-value

$(\sigma_v')_n$: standard effective overburden pressure = 0.6kgf/cm²

Moreover, the relationships between SPT N-values and the terms, Tn, were summarized by the steps of 20 % in fine contents.

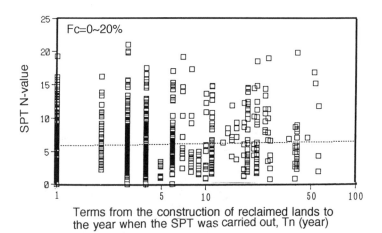

Fig.15 Relationships between Tn and SPT N-value for 0 to 20 % in fine contents

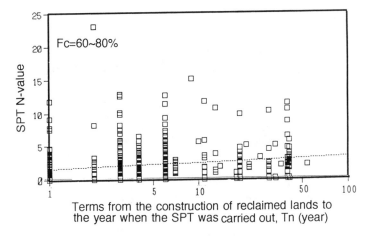

Fig.16 Relationships between Tn and SPT N-value for 60 to 80 %
in fine contents

Figure 15 and Fig.16 show the relationships of SPT N-values and the
terms, Tn, of 0 to 20 % in fine contents and of 60 to 80 % in fine contents,
respectively.   Though the data are faily scattered, SPT N-values have
tendencies to increase with time. Broken lines in the figures show average
relationships. Figure 17 shows relationships between fine contents and the
increments of SPT N-values from one year after reclamation to 50 years
after reclamation, ΔN_{50}, read from the average relationships shown in Fig.15
and Fig.16. As the increment of SPT N-values with time, ΔN_{50}, increased
with fine contents, it can be said that the aging effect on SPT N-values also
becomes strong with contents of fine particles.

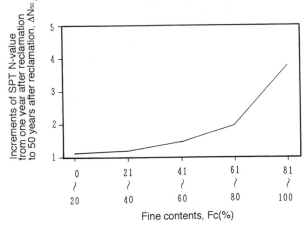

Fig.17 Relationships between Fc and ΔN_{50}

Conclusions

Several undrained cyclic triaxial tests on undisturbed and reconstituted samples, and case studies on liquefaction histories of artificailly reclaimed grounds, were conducted to demonstrate the liquefaction strength of artificially filled silty sands. From the studies mentioned above, the following conclusions can be drawn.

(1) The liquefaction strength of a silty sand was not large and not very different from that of a fine sand just after reclamation. However, the liquefaction strength of a sand containing many fine particles increases faster than the liquefaction strength of a fine sand.
(2) The minimum peak surface acceleration which causes liquefaction at reclaimed lands of silty sands increases with time faster than the minimum peak acceleration at reclaimed lands of clean sands.
(3) SPT N-value of the artificially filled silty sand increases with age faster than the SPT N-value of the artificially filled clean sand.

Acknowledgement

The authors would like to express their thanks to Mr.T.Miyamoto of Nippon Steel Co., Mr.K.Koga of Asahi Chemical Co., Mr.T.Yoshihara of Mitsubishi Estate Co., Ltd. and Mr.T.Yanagihata of Kyushu Institute of Technology for their assistance in the study.

References

Kimura,M., Tatsuoka,F. and Pradhan,T.B.S (1986): Liquefaction Resistance of Two Different Sand Under Long Term Consolidation and Over-Consolidation Stages. Proc. of the 21th Japan National Conf. on SMFE, pp.591-594 (in Japanese).
Troncoso,J., Ishihara,K. and Verdugo (1988): Aging Effects on Cyclic Shear Strength of Tailing Materials. Proc. of 9th WCEE, Vol.III, pp.121-126.
Yasuda,S., Tohno,I., Morimoto,I. and Miyamoto,T.(1989): Liquefaction of Reclaimed Land during the 1987 Chibakentoho-oki Earthquake. Proc. of the 4th Int. Conf. on Soil Dynamics and Earthquake Eng., pp.27-36.
Yasuda,S., Nagase,H., Yanagihata,T.(1993): Effects of Fine Contents and Aging on the Liquefaction Strength of Artificially Filled Silty Sands., Performance of Ground and Soil Structures during Earthquakes (Special Volume for 13th ICSMFE),pp.159-164.
Wakamatsu,K., Yasuda,S., Yoshida,N. and Yoshihara,T.(1992): Liquefaction History of Reclaimed Land. Proc. of the 27th Japan National Conf. on SMFE, pp.1063-1066 (in Japanese).

Liquefaction Characteristics of Silts

Sukhmander Singh[1]

Abstract

Cyclic triaxial load testing on laboratory constituted and undisturbed samples of silt and silty sands were made to study the liquefaction characteristics of silts. A systematic study on the cyclic load testing of laboratory prepared samples of silts and sands with 10, 20, 30 and 60 percent silt by weight indicate that sands containing 10, 20 or 30 percent silt have lesser resistance to liquefaction than 100 percent sands. This is so if comparison is made for samples prepared at the same relative density of about 50 percent. Interpretation of results in terms of void ratio does not hold to the same conclusion. The test data on laboratory constituted and undisturbed samples of silts, points out the difficulty in establishing liquefaction criteria for undisturbed deposits of silts in terms of joint pore pressure and deformation characteristics. Unlike sands, and lab constituted samples of silts, undisturbed samples of silts develop strains before a significant pore pressure increase is recorded.

Introduction

In connection with the cyclic triaxial testing of undisturbed samples of silts from various projects (Ref. 2, 3, and 4), it became apparent that the pore pressure generation and strain development characteristics of the silty samples were quite different than that of sandy samples. Attempts were made to examine closely the differences in the liquefaction characteristics of sands and silts. In the past 25 years the phenomenon of the liquefaction of sand deposits has been the subject of intensive research. As a result, significant improvement has occurred in our ability to measure and predict the probable performance of a wide range of sandy soils during an earthquake. However, as the literature was

---

[1]Chairman, Department of Civil Engineering, Santa Clara University, Santa Clara, California, 95050

searched a noticeable gap was found to exist when the soils subjected to shaking are silts and sandy silts. Especially, comparative studies on the cyclic strength behavior of reconstituted and undisturbed samples of silts were rare or non-existent. Accordingly, the research described herein was undertaken to study liquefaction characteristics of silts. The approach adopted was to study systematically the cyclic strength characteristics of laboratory prepared samples of sands with different percentages of silts and then to test 100 percent silt samples to compare with test results on undisturbed samples of silts and silty sands. The automated cyclic triaxial system was used to test samples in a controlled laboratory environment. The following sections present the test results first followed by discussions.

Test Results on Undisturbed Samples

Figure 1 shows the results of a cyclic triaxial test on a typical silt sample from Valdez. The upper curve shows the development of strain with increasing numbers of cycles and the lower curve shows the rate of pore water pressure increase. It is readily apparent that the increase in both strain and pore water pressures takes place gradually. This is quite different from the behavior of sands (Seed and Lee, 1966; Finn et al, 1970; Singh et al, 1980; and Finn, 1981) where strains which had previously been small start to increase rapidly as the pore water pressure approaches 60 percent or more of the confining pressure.

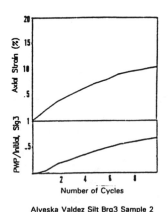

Alyeska Valdez Silt Brg3 Sample 2

Figure 1.

Figure 2 shows the peak-to-peak axial strain in percent plotted against the pore

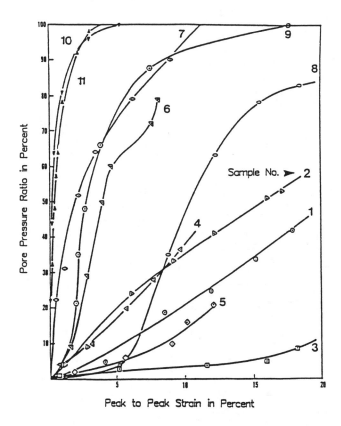

Figure 2.

water pressure increase, also expressed as a percentage (pore pressure ratio equals 100 percent when the pore water pressure equals the confining pressure) for a wide range of samples. For most of the silt samples (samples #1, 2, 3, 4, 5) whose results are shown in figure 2 large strains are reached without the development of significant pore pressures. The exception appears to occur for loose silt (sample #6) which approximates the curves for sand (sample #7, 9, and 11) shown on Figure 2. The envelope curves for the grain size of the soils whose test results are given in Figure 2 are shown in **Figure 3**.

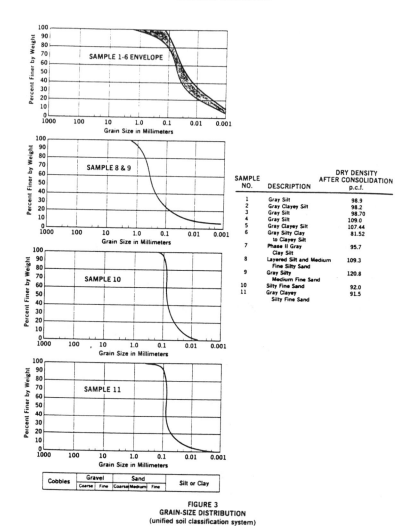

FIGURE 3
GRAIN-SIZE DISTRIBUTION
(unified soil classification system)

Test Results on Laboratory Prepared Samples

The following sections present a summary of each of the test series carried out on laboratory constituted samples and the associated results.

The first test series included setting the base line curves for clean sands. Two types of sands, Flint Shot No. 4 and Standard Ottawa Sand were tested under cyclic loadings for relative densities of 50 percent. The base line curve describes in a conventional way the relationships between cyclic stress ratio and number of cycles. The grain size distribution curves are shown in Fig. 4.

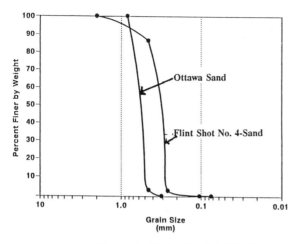

Figure 4. Grain Size Distribution

The base line curve for Flint Shot No. 4 sand, which was chosen as the basis for comparison, is shown in Fig. 5. Uniform silt was obtained from sieving of fine sand obtained from Pasadena California.

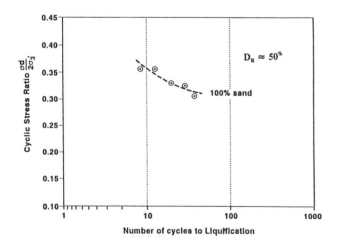

Figure 5. Cyclic Stress Ratio vs No. of Cycles to Initial Liquefaction

The second set of test series involved cyclic load testing of sand samples, each containing 10% silt by weight. Trials were first made to prepare a uniform density sample with a relative density of 50%. Moist tamping method was selected. Each sample was prepared in six layers. The moisture content for the sample was chosen after some trial. Typically 7 to 12% moisture content was found to be appropriate. Each sample was prepared at approximately 50% relative density.

The third and fourth test series were respectively carried out for 20% and 30% of silt content by weight in sand samples. The sample preparation techniques and the testing procedures were similar to that of the second test series. The plots for the second, third and fourth test series are shown in Fig. 6. Each sample was prepared to approximately 50% relative density.

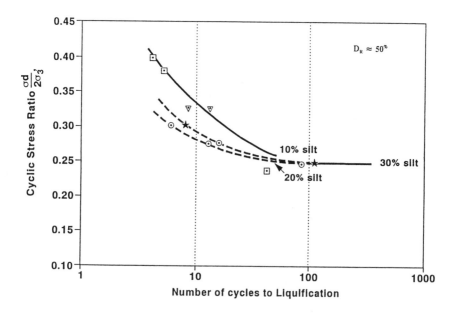

Figure 6. Cyclic Stress Ratio vs No. of Cycles to Initial Liquefaction

Sample preparation for a 100% silt sample was a challenge. The difficulty was in choosing right moisture and achieving the right density. However, after some trials, this was resolved and a curve describing the relationship between cyclic stress ratio and number of cycles was obtained and is shown in Fig. 7. Again, a relative density of about 50% was achieved for each of the silt samples.

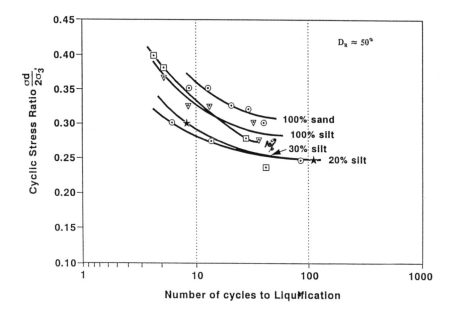

Figure 7. Cyclic Stress Ratio vs No. of Cycles to Initial Liquefaction

In all of the test series involving silt, it was found that the results were very sensitive to the method of sample preparation. Consistancies in all aspects such as mising, tamping of each layer and handling was very important to achieve consistant results. The presence of a somewhat weaker layer was readily evident from the manner in which the sample would fail.

Discussions

Undisturbed Samples

Although pore pressure and deformation characteristics under cyclic loading are closely related, there is a problem in establishing a liquefaction criteria for silts in terms of joint pore pressure and deformation characteristics. In almost all of the undisturbed silt samples tested, a gradual softening of the sample was observed from the start of the test. **Figure 8** shows a comparison of pore pressure development in a typical silt sample and a typical loose sand sample. For most of the silt samples tested, the trend of pore pressure development was different than for sands. It takes a large number of cycles of loading to reach

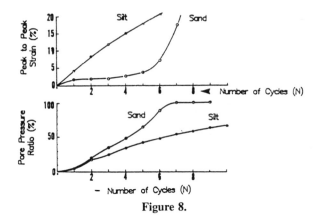

Figure 8.

the 100 percent pore pressure ratio which is similar to the buildup of pore pressures in a dense sand sample. Zhaoji (1988) observed similar behavior for the undisturbed samples of silt during cyclic triaxial liquefaction tests. As also pointed out by Zhaoji (1988), it appears that the undisturbed samples of silt develop a structure due to cementing at grain contacts during the long geologic history; and this affects the liquefaction susceptibility and pore pressure generation characteristics of undisturbed silts. Studies by Parkash and Puri (1982) on the liquefaction of loessial soils also indicate that the cohesion component of the undisturbed loessial soils delays the development of pore pressure, and that in almost all of the tests, the axial strains in the range of 10% double amplitude axial strain, or more, had developed by the time pore pressures became equal to effective confining pressure. Clearly the importance of structure in fine grained soils such as silts need to be recognized when evaluating the pore pressure generation and strain developments of these deposits.   In some of the silt samples a pore pressure ratio of 100 percent was never reached.In cases where deformations reach a significant magnitude well before the pore pressure reaches 100 percent, the liquefaction and the dynamic strength loss characteristics of silts may be defined only in terms of the percent of strain irrespective of the magnitude of the pore pressure. For loose silts where the 100 percent pore pressure ratio was reached and accompanied by large strains, the liquefaction criteria can be described by a 100 percent pore pressure ratio with 10 percent of strains potential. This is similar to the criteria presently used for sands.

<u>Reconstitued Samples</u>

It may be noted from Fig. 3, that sands containing 10, 20, or 30% of silt by weight have lesser cyclic strength than that of 100% sands. These samples were prepared to the same relative density of 50%. Maximum and minimum densities were determined separately for each case, and 50% relative density estimated from the maximum and minimum densities. It has been suggested that relative density is not a suitable index for characterizing behavior of silty sands (Ishihara et al, 1980). Laboratory controlled studies on the effect of silt content on cyclic strengths of clean sands have been reported by Chang et al (1982) and Kuerbis et al, 1988). Chang et al used void ratio as the basis for comparison and tested silty sand samples at the same void ratio as the 100 percent sand samples, and concluded that the cyclic shear resistance increases as the silt content increases; that the increase is slight, up to 10% silt content, and begins to develop more as the silt content increases to 30 percent silt. Chang et al explains this behavior in terms of sand grain to sand grain contact in the soil structure. But it appears to be due to the increase in relative density as the more silt is added and the sample is prepared at the same void ratio. On the other hand, studies by Kuerbis et al (1988) suggest that the higher the silt content, the lower the cyclic resistance for a given relative density. Studies reported herein indicate similar results (figure 7). Kuerbis et al also reported that over the range of overlapping void ratios of various silty sands, the cyclic resistance at a given void ratio decreases with increase in silt content. Kuerbis et al used the concept of sand-skeleton void ratio to explain the results and concluded that the rather small change in behavior with increasing silt can be explained by the fact that the sand skeleton void ratio remains virtually unaltered with the addition of silt. Silt merely acts as a filler of sand skeleton voids and hence behaves as an inert component (up to about 20% of silt). Tokimatsu and Yoshiji (1984) and Seed et al (1985) suggested on the basis of field performance of sandy soils with fines that soils containing more than 20% clay (finer than $5\mu m$) would hardly liquefy unless their plasticity indexes are low.

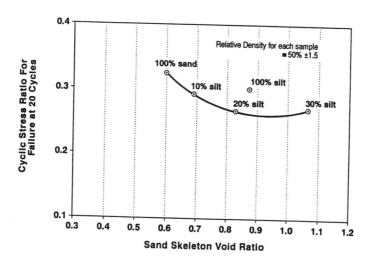

Figure 9. Cyclic Stress Ratio vs Sand Skelton Void Ratio

From the foregoing, it appears that it is not easy to clarify the influence of fines on the behavior of sands under cyclic loading, and that the liquefaction characteristics of silts may not be estimated on the basis of criteria currently used for sands. In an attempt to explain the behavior in terms of void ratio, figure 9 was prepared; and it appears that whereas, void ratio may be the better indices than relative density, to explain the influence of fines on sand behavior, the

interpretation in terms of void ratio alone beyond a percentage of fines of 60% (figure 10) need further studies and should be made with caution.

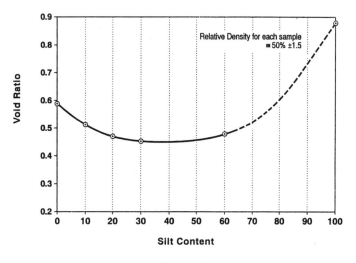

Figure 10.

Conclusions

On the basis of the studies reported herein, the following conclusions may be drawn:

1.   There is a difference in the strain developments and the pore pressure generation characteristics of undisturbed and laboratory prepared samples of silts tested under cyclic triaxial loading conditions.

2.   There is a problem in establishing criteria for liquefaction of undisturbed silts in terms of joint consideration of pore pressure and deformation.

3.   For sands and laboratory prepared samples of silts, it is possible to use pore pressure criteria to describe terms such as initial liquefaction and

estimated strain potential. For most undisturbed silts this is not possible because the 100 percent pore pressure increase is not reached during testing. The strains develop before a significant pore pressure increase is recorded.

4. The structure, and hence the deposition of a silt deposit is an important parameter which appears to slow down the pore pressure generation, but may have the opposite effect on the strain development and needs further study.

5. For a given relative density, sands containing 10, 20 or 30% of silt by weight have lesser resistance to liquefaction than 100% sands. The loss of resistance appears to be consistant when results are compared in terms of void ratio, thus implying that void ratio may be a better indices to use for silty fine sands and silts.

6. 100% silt samples tested at the same relative density as the 10, 20, and 30% silt in sand showed higher strengths, but not higher than 100% sands.

Acknowledgements

The research reported herein was supported through the National Science Foundation Grant No. 9013422, and is gratefully acknowledged. Dr. David G. Toll of Durham University, is acknowledged for his constructive suggestions. Manjula R. Mannem meticulously performed the cyclic triaxial testing and assisted in the research. Her efforts are highly appreciated.

References

1. Chang, N.Y., S.I. Yeh, and L.P. Kaufman, (1982), "Liquefaction Potential of Clean and Silty Sands," Third International Earthquake Microzonation Conference, Proceedings, Seattle, Washington, Vol. II, pp. 1017-1032.

2. Dames & Moore, (1975), Outfall Diffuser and Dike Stability Studies, Valdez Terminal, D&M Project 8354-059-20.

3. Dames & Moore, (1980), Laboratory Dynamic Soil Testing, Prez Caldera No. 1 Tailings Dam, Chile, D&M Project 10438-003-03.

4. Dames & Moore, (1982), Geotechnical Studies, Klaune Lake Crossing, Robinson-Dames & Moore.

116	GROUND FAILURES

5. Finn, W.D., L. Liam, L. Bransby and D.J. Pickering, (1970), "Effect of Strain History on Liquefaction of Sand," Journal of the Soil Mech. and Found. Div. ASCE, No. 96, No. SM6, Nov.

6. Kuerbis, R., D. Nequssey, and Y.P. Vaid, (1988), "Effect of Gradation and Fines Content on the Undrained Response of Sand," Proceedings, of the ASCE Geotechnical Division Specialty Conference on Hydraulic Fill Structures, Colorado State University, Fort Collins, Colorado, August.

7. Prakash, S. and V.K. Puri, (1982), "Liquefaction of Loessial Soils," Third International Earthquake Microzonation Conference, Proceedings, Seattle, Washington, Vol. II, pp. 1101-1107.

8. Seed, H.B. and K.L. Lee, (1966), "Liquefaction of Saturated Sands During Cyclic Loading," Journal of Soil Mechanics and Foundations, ASCE, Vol. 92, No. SM6, Proc. Paper 4972.

9. Seed, H.B., K. Tokimatsu, L.F. Harder, and R.M Chung, (1985), "Influence of SPT Procedures in Soil Liquefaction Resistance Evaluations," Journal of Geotechnical Engineering Division of ASCE, Vol. III, No. 12, December.

10. Singh, S., N.C. Donovan, and T. Park, (1980), "A Re-examination of the Effects of Prior Loading on Liquefaction of Sands," Proceedings, Seventh World Conference on Earthquake Engineering. Istanbul, Turkey, Aug., 1980.

11. Singh, S., and R.Y. Chew, (1988), "Dynamic Strain in Silt and Effect on Ground Motions," Paper presented and published at American Society of Civil Engineers Division of the Geotechnical Engineering's Earthquake Engineering and Soil Dynamics II Conference, Park City, Utah, June 27-30.

12. Tokimatsu, K., and H. Yoshimi, (1981), "Field Correlation of Soil Liquefaction with SPT and Grain Size," Proceedings, of the International conference on Recent Advances in Geotechnical Earthquake Engineering and Soil Dynamics.

13. Tokimatsu, K., and ?. Yoshimi, (1984), "Criteria of Soil Liquefaction with SPT and Fines Content," Proceedings, of the 8th World Conference on Earthquake Engineering, Vol. III, San Francisco, California, July.

14. Zhaoji, S., Y. Shousong, W. Yuging, and Y. Shihong, (1984), "Prediction of Liquefaction Potential of Saturated Clayey Silt," Earthquake Engineering and Earthquake Vibration, China, September.

15. Zhaoji, S., (1988) "Study of Silt Liquefaction During Earthquakes," Institute of Engineering Mechanics, SSB Harbin, China.

BEHAVIOR OF GRAVELLY SOILS DURING SEISMIC CONDITIONS - AN OVERVIEW

Presentation Summary

M. E. Hynes, M. ASCE[1]

INTRODUCTION

INTRODUCTION

Gravelly soils continue to challenge our creative engineering abilities in seismic loading problems. Although the number of case histories of seismically-induced liquefaction of gravelly soils is growing, the empirical data base is uncomfortably sparse for establishing simplified design charts. The papers collected for this session and other recent publications provide a wealth of new data from case histories and investigations of gravelly soils which improve our understanding of the behavior of gravelly soils subjected to seismic loading and refine our seismic design procedures. The contributions of the authors for this session are discussed in this overview.

FIELD OBSERVATIONS OF LIQUEFACTION OF GRAVELLY SOILS

Valera et al. examine the characteristics of seven reported cases of liquefaction of gravels, namely the liquefaction of a gravelly alluvial fan during the 1948 Fuqui earthquake in Japan; a flow slide at Valdez and bridge foundation behavior in the 1964 Alaska earthquake; slides in Shimen Dam during the 1975 Hanking and Baihe Dam during the 1976 Tangshan earthquakes in China; and liquefaction at Pence Ranch and Whiskey Springs during the 1983 Borah Peak, Idaho earthquake. Andrus (1994, PhD Thesis at Univ. of Texas at Austin) further analyzed Goddard Ranch, Andersen Bar, and Larter Ranch, more of the affected 1983 Idaho

---

[1] Chief, Earthquake Engineering and Seismology Branch, Geotechnical Laboratory, USAE Waterways Experiment Station, Vicksburg MS 39180-6199

sites. Lum and Yan, and Harder remind us of two non-
liquefied gravel sites, Mackay Dam and North Gravel
Bar, in the 1983 Borah Peak, Idaho earthquake. Yegian
et al. (1994 ASCE-GT3) reports liquefaction of two
gravelly sites during the 1988 Armenia earthquake, as
well as one non-liquefied site; Yegian et al. (1994)
also report an estimate of residual strength of 100-260
psf for a gravel with SPT blowcounts of about 12
blows/ft. Jamiolkowsky et al. (1992 Wroth Memorial
Symposium) reports evidence of liquefaction in the
recent Messina gravels, discovered in the field
investigations for the proposed bridge over the Straits
of Messina. Kokusho and Tanaka introduce yet another
observation of liquefaction in a gravelly volcanic
debris deposit during the 1993 Hokkaido-Nanseioki
earthquake.

The contributions of all the authors in this session
provide a basis for detecting common patterns from
these case histories that are generally consistent with
the conclusions of Valera et al.:

*   There is ample evidence that liquefaction of loose
    to medium dense gravelly deposits can and does
    occur.

*   The drainage and dissipation characteristics of the
    deposit affect the potential for liquefaction and
    the extent of resulting damage. For example, the
    presence of fine-grained layers that impede
    drainage can result in liquefaction of a deposit
    that would not liquefy if the aquaclude were not
    present (from Yegian et al. 1994, ASCE-GT3).

*   Penetration testing of gravelly sites are presently
    our most practical approach for characterizing the
    seismic resistance of these sites in situ.
    Analyses of the case histories continue to indicate
    that simplified empirical liquefaction procedures
    (e.g., Seed et al. 1985) are generally appropriate
    for evaluating liquefaction susceptibility of
    gravelly sites.

FIELD INVESTIGATION OF GRAVELLY SITES
FOR SEISMIC EVALUATION

The presence of gravel, cobbles and even boulders
limits the applicability of field investigation
techniques developed and refined for sands and clayey
soils. Procedures for investigating gravelly soils and
interpreting results are developing to obtain better
undisturbed samples, measure in situ density, perform

geophysical tests, and measure penetration resistance.
These advances are demonstrated by the contributors to
this session: Kokusho and Tanaka, Tonno et al., and
Goto et al. report studies involving ground freezing to
obtain high quality undisturbed samples of gravels and
subsequent laboratory tests for comparison with field
measurements including shear wave velocities and
penetration resistance using Standard and Large-scale
hammers (Jamiolkowsky, as cited above, also describes
the calibration and use of a large-scale penetration
hammer). Lum and Yan provide a study of various
methods of performing and interpreting penetration
tests, including Standard and Becker penetration tests,
and shed light on Harder and Sy methods of interpreting
Becker penetration tests. Valera et al. suggest that
for most practical applications, SPT is sufficient for
evaluating gravelly sites. Tonno et al. report the
results of extremely large-scale tests on a gravelly
site to determine adequacy of such sites as foundations
for nuclear power plants.

A generalization that can be drawn from these studies,
emphasized by Lum and Yan and Kokusho and Tanaka, is
that there continues to exist a large scatter in
correlations of shear wave velocities in gravelly soils
to other parameters such as penetration resistance or
void ratio, although fairly consistent relationships
can be developed for individual sites.

LABORATORY TESTING OF GRAVELLY SOILS FOR SEISMIC EVALUATION

Kokusho and Tanaka, Tonno et al., and Goto et al.
report the results of large scale cyclic triaxial tests
on high quality undisturbed (frozen in situ) samples
and reconstituted samples. As expected from past work
on sands, the undisturbed specimens had much higher
low-strain moduli (G_o) and higher cyclic strengths than
the reconstituted samples. However, these studies also
demonstrated that the relative behavior of the gravelly
soil was well represented by the reconstituted
specimens--the ratio of G/G_o versus shear strain was
nearly identical for the undisturbed and reconstituted
specimens.

Evans and Zhou present the results of cyclic triaxial
tests on soils with increasing percentages of gravel
content and conclude that the presence of the gravel,
even with only floating particles in a finer soil
matrix, increases the liquefaction resistance of the
mixture. This result raises additional questions about

the relationship between laboratory test results and
actual field behavior.

RESIDUAL UNCERTAINTIES AND FUTURE DIRECTIONS

Experience in the field and in the laboratory with
gravelly soils and closer examination of case histories
has improved our understanding of their behavior under
seismic loading. But several issues remain. In the
field, we have yet to perfect a low cost, fully
understood, three-dimensional procedure for mapping
deposit stratigraphy and measuring directly the
material properties needed for seismic analysis, but we
are getting closer. In our understanding of potential
for volume change of gravelly soils, we have yet to
develop a procedure to correctly, exactly determine the
minimum and maximum void ratios for a given gradation.
Advances here may come from numerical packing studies
or from in situ compaction procedures, or elsewhere.
Membrane compliance continues to cast uncertainty on
the behavior of laboratory tests and their
applicability to the field. Without doubt, our
understanding of these materials will continue to
improve as more sites are investigated, more case
histories are analyzed, and new laboratory methods such
as centrifuge testing are exercised.

Dynamic Properties of Gravel   Layers
Investigated by
In-Situ Freezing Sampling

Takeji Kokusho[1] and Yukihisa Tanaka[2]

Abstract

        Dynamic properties of Pleistocene gravel layers
are investigated by means of in-situ freezing sampling and
cyclic triaxial tests. The small-strain modulus, the
strain-dependency of modulus, the hysteretic damping and
the undrained cyclic strength are measured for a number of
high-quality intact specimens sampled by the freezing
method from four different sites. These properties are
correlated with other soil parameters to propose empirical
formulas to estimate the in-situ dynamic properties of
Pleistocene gravelly soils.

Introduction

        Gravel layers are widespread and often encountered
in civil constructions. Deeper parts of plain areas
sometimes consist of gravels of Pleistocene epoch, and
fluvial river beds are covered with gravels of younger
ages. These gravel layers are normally of high density
and serve as bearing strata for pile foundations. Recently
they are increasingly serving as bearing strata for direct
foundations of important structures such as high-rise
buildings, suspension-bridge piers, nuclear power facilities,
rock-fill dams, etc.

        Because of their high densities these gravel
layers are normally believed seismically stable without
posing any significant problem in seismic design. On the
other hand there have been some case histories in which
presumably loose gravel soils did liquefy and boiled out

1.Senior Research Fellow, 2.Principal Research Engineer
  Central Research Institute of Electric Power Industry
  (CRIEPI) 1646, Abiko, Abikoshi, Chiba-Ken, JAPAN

121

(e.g.   Andrus et al. 1987, Ishihara et al. 1989).   One
of the most typical cases of this kind has been found very
recently during the Hokkaido-Nanseioki Earthquake(M=7.8)
in 1993 in which volcanic debris containing large amount
of gravels liquefied causing damages of a number of houses
due to differential settlement (Kokusho et al. 1994).
These cases clearly indicate that even gravelly soils if
loose enough are susceptible to liquefaction as loose
sands.   With increasing construction projects of important
structures on gravel layers, it is of utmost importance to
investigate their in-situ dynamic properties which has
been least focused on so far.

        Significant difficulties arise when one tries to
apply conventional soil investigation techniques either of
in-situ testing or of soil sampling to gravelly soils due
to the existence of large grain particles.   Normal soil
sampling methods are almost inapplicable to gravels and
even large diameter samplers available for this purpose
are almost powerless to recover intact samples for the
purpose of evaluating dynamic properties.   The importance
of intactness of cohesionless soils has been evidenced in
previous researches(e.g. Kokusho,1987) in which the undrained
cyclic strength of sands was found to considerably decrease
due to minimal mechanical disturbance during sampling
procedures.   The denser the sand is, the more pronounced
the decrease is.   It is therefore indispensable to
recover gravelly soil samples as intact as possible to
reliably evaluate in-situ dynamic properties of gravel
layers if one wants to rationalize seismic designs for
foundations on gravels.

        For this purpose the authors have developed
in-situ freezing sampling method and applied it to gravel
layers of several different sites.   In the first part of
this paper the in-situ freezing sampling technique is
briefly described and then fundamental soil properties
such as in-situ void ratio, particle size distribution
etc. as measured based on the intact samples are discussed.
In the latter part, laboratory test results based on the
intact samples are described in terms of the shear modulus,
the hysteretic damping and the undrained cyclic strength.
These values are compared with other soil parameters such
as the void ratio, the modulus, the penetration resistance,
etc., and their correlations are discussed to establish
some empirical formulas for simplified evaluations of
in-situ dynamic properties of gravels.

In-Situ Freezing Sampling Method

        A special sampling technique has been developed

for obtaining gravels as intact as possible by first freezing gravels and then coring it out by a special sampler. The technique has been applied to several different sites and improved step by step.

In this sampling method, double-tube freezing pipes are first installed by drilling a gravel layer as illustrated in Fig.1 with great care not to give large disturbance to the surrounding soil. At the same time large diameter holes are drilled and casings are set in the upper soil layer for subsequent sampling procedures in the underlying gravel layer. Then the liquid nitrogen (LN2) of -196°C in temperature is supplied into the inner tube of the central freezing pipe and evaporated at the lower part of the outer tube taking out the heat of the surrounding soil. The evaporated gas coming out of the central pipe still cold enough (-130°C) is lead to neighboring pipes, eventually making a

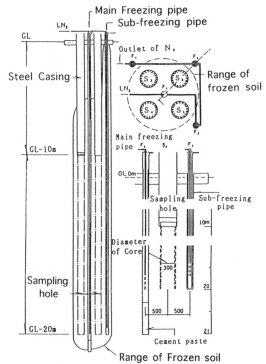

Fig.1 Set-Up for In-Situ Freezing Sampling

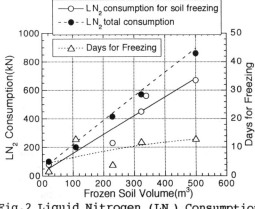

Fig.2 Liquid Nitrogen (LN$_2$) Consumption and Freezing Days versus Frozen Soil Volume

frozen soil column of about 2m in diameter. Temperature changes during freezing are measured with gauges in various points in the gravel layer. Based on preliminary researches, the threshold temperature was chosen as -10°C and the frozen-soil coring was started after the soil temperature for the cored soil was lowered below this threshold value. Fig.2 shows the consumption of the liquid nitrogen and the number of days plotted against the volume of the frozen soil column based on the actual experiences in several sites.

Two kinds of sampler have been developed for recovering cored frozen soils; 10cm/30cm inner diameter samplers. Both belong to the triple-tube sampler consisting of the outer tube with diamond bit, the intermediate tube for accommodating the sample intact and the inner-most split tube for easy pull-out of sample. Soil coring starts into the frozen soil mass from the bottom of the previously drilled holes protected by casings.

Fig.3 Sampled Frozen Gravel in 30 cm-Diameter Sampler

Either the calcium chloride (CaCl$_2$) or the ethylene glycol (HOCH$_2$CH$_2$OH) is used as the mud fluid for cooling the drilling bit and removing the tailings. The temperature of the fluid must be regulated at about -10°C just above the freezing point of the fluid.

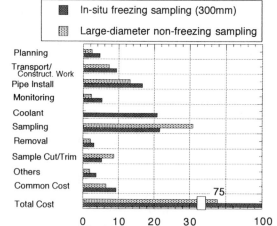

Fig.4 Cost Break-Down of In-Situ Freezing Sampling and Non-Freezing Sampling

Fig.3 shows the picture of cored gravel   still in the sampler of 30cm in diameter.

Based on the experiences, the cost evaluation of the in-situ freezing sampling has been carried out . Fig.4 shows the comparison of costs in two sampling methods; in-situ freezing sampling and non-freezing large diameter sampling for taking gravel samples of 30cm in diameter under the T-site condition as shown in Fig.5. This typically indicates the non-freezing sampling will be 25% cheaper than the freezing one if it would be worthwhile to apply to gravels.   The principal cause for the higher cost for the freezing one is of course the necessity of the coolant (the liquid nitrogen).

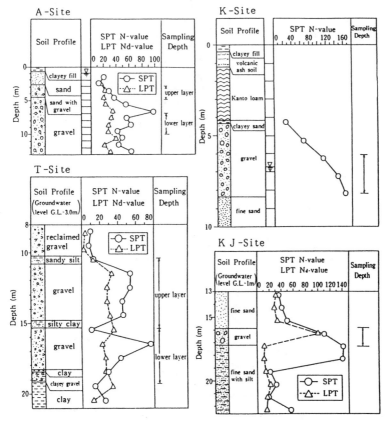

Fig.5 Soil Profiles at Four Gravelly Sites

Test name	SPT: Standard Penetration test	LPT: Large Penetration Test
Weight(N)	622.3	980
Drop height (cm)	75	150
Drive length (cm)	30	30
Details of Penetration Probe	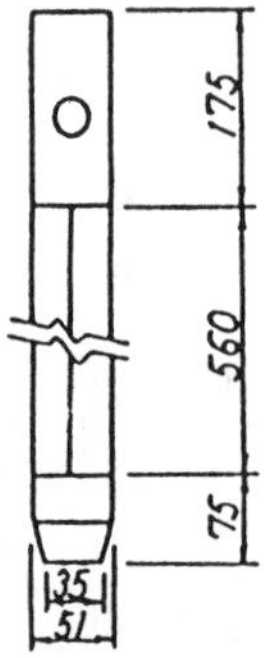	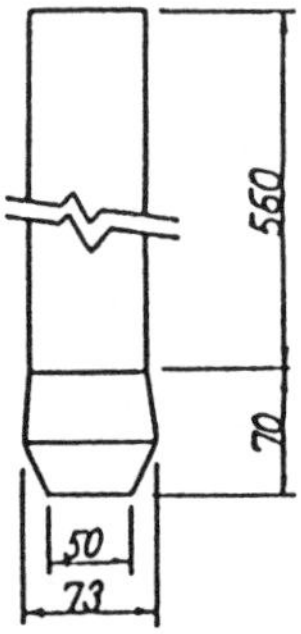

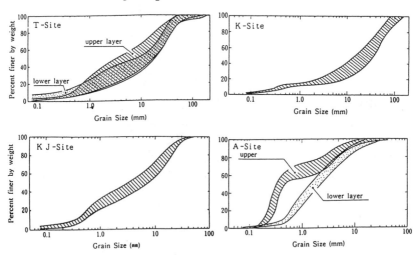

Fig.6 Specifications of SPT and LPT

Fig.7 Grain Size Distributions of Gravels at Four Sites

Soil   Profile   and
Physical Properties

    In   four
different sites, named
T, KJ, A and K here,
the sampling has been
carried out (Tanaka et
al. 1988,1990,1991).
Fig.5 shows the soil
profiles   and   the
penetration resistance
as well as the sampling
depth in the four sites.
Two kind of penetration
tests have been carried
out here; the standard
penetration test (SPT)
and   the   large
penetration test (LPT),
the specification of
which is available in
Fig.6.   In LPT, the
driving energy is 3.15
times larger than SPT,
allowing   smoother
penetration   into
gravels with coarse
grains (Yoshida and
Kokusho 1988).   Fig.7
illustrates   the
particle   size
distribution of all the
samples taken in the
four sites.   All gravel
layers are from the
Pleistocene epoch and
contain only a small
amount   of   fines.
Fig.8a,b,c shows the
void   ratios,   e,
evaluated from the
frozen gravel samples
in the four sites
plotted against the
uniformity coefficient,
Uc, the gravel content,
Gc, and the fine
content, Fc. The figure
clearly indicates the
in-situ void ratio of
gravelly soils has a

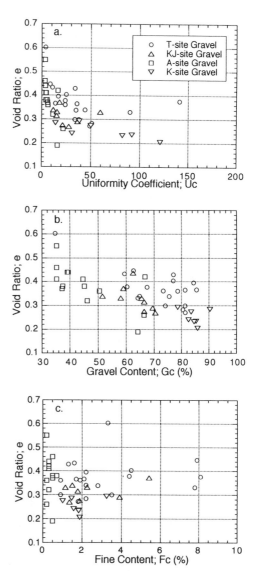

Fig.8 In-Situ Void Ratio
      Plotted against
      Uniformity Coefficient(a),
      Gravel Content(b) and
      Fine Content(c)

strong dependency on the grain size distribution.
Namely, the void ratio decreases with the increase of the
uniformity coefficient or the gravel content.   It is
also noted that the increase in the fine content has a
reverse effect on the void ratio despite the increase in
Uc.   Anyway it is obvious that the void ratio in typical
Pleistocene gravel layers attain as low as 0.2 to 0.3.
One of the significant features of gravel layers in
contrast to sand layers is this low void ratio (high
density) which may result in higher stiffness and higher
strength.

In Fig.9 the maximum and minimum void ratio, e_{max}
and e_{min}, as well as in-situ void ratio, e, are shown
against the uniformity coefficient, Uc, for all specimens
taken from the two sites where these data are available.
The minimum void ratio, e_{min}, here was determined by using
a large-scale mold and a   vibrating cap shown in Fig.10b.
Ikemi   et   al.   (1984)   presented the relationship as
shown in Fig.10a between the dry density and the uniformity
coefficient of four kinds of artificially mixed gravels
and the Toyoura standard sand, indicating that; (1)this
method can attain the highest dry density among other
methods including the ASTM method using the vertical
shaking table, and (2)the dry density attained by this
method is almost equivalent to that by the method standardized
by the Japan Society for SMFE.   On the other hand the

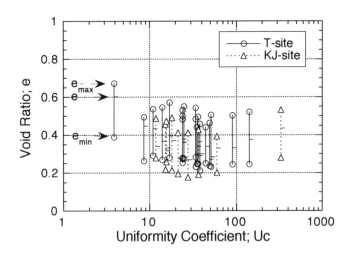

Fig.9 Maximum, Minimum and In-Situ Void Ratio(e_{max}, e_{min}, e)
     vs Uniformity Coefficient

maximum void ratio, e_{max}, here are determined with basically
the same method as others by gently filling the same-size
mold by a shovel.

It is interesting to find out that the three kinds
of void ratio tend to decrease with increasing uniformity
coefficient up to Uc=100 and then again increase probably
due to the increase in the fine content.   It is also
noted that the difference in void ratio, $e_{max}-e_{min}$, is
almost constant despite the wide variety of the uniformity
coefficient.   This fact is also confirmed in Fig.11 where
$e_{max}-e_{min}$ is plotted against the mean grain size, D_{50}.   All
the gravels take the value around 0.2 to 0.3 whereas sands
as shown by Ishihara(1976) exhibit much higher values.
This indicates gravels are less susceptible to liquefaction
than sands with the same relative density, because the
room for volumetric strain, $(e-e_{min})/(1+e)$, is considered
smaller for gravels than sands.

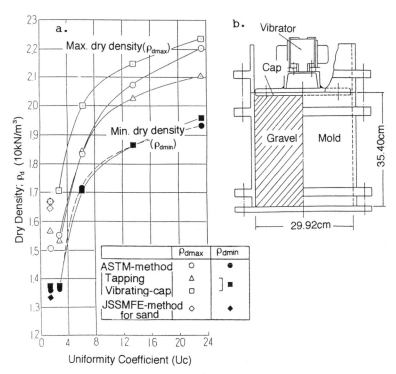

Fig.10 Max. and Min. Density Test
       a.Typical Results for Gravels and Sands
       b.Test Mold

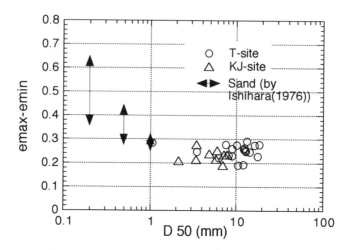

Fig.11 Difference in Void Ratio($e_{max}-e_{min}$)
vs Mean Grain Size(D_{50})

Laboratory Mechanical Tests

Test Method

Sampled gravels, either of 30cm(T, KJ, K-site) or
10cm(A-site) in diameter, were set up in the triaxial
apparatus in still frozen condition and, under an isotropic
confining stress, gradually thawed.   Then the specimen
was completely saturated with the aid of the carbon
dioxide gas.   Specimens were basically tested under the
isotropic effective confining stress equal to the in-situ
effective overburden stress.

In order to reduce the membrane penetration effect,
the side surface of specimens was smoothed except those
which already had a smooth enough side surface due to rich
containment of fine sand particles.   The smoothing was
done by patching fine sand on the frozen specimen and
trimming again to obtain a smooth surface (Tanaka, Kokusho
et al. 1991).

Two kinds of cyclic triaxial tests were carried
out; the cyclic deformation test to obtain the shear
modulus and the hysteretic damping   for small to medium
strain level and the undrained cyclic strength test for
large strain.

In order to make reliable measurement of the
modulus and the damping for a small strain level, the

triaxial apparatus employs the internal load cell just
above the loading plate and non-contact type high-sensitivity
deformation gages in the triaxial chamber (Kokusho et al.
1983, 1987). The continuous variation of the modulus and
the damping can be consistently measured for a wide strain
range of 10^{-6} to 10^{-3} with this improved system. For all
specimens the small-strain shear modulus for a strain
level of 10^{-6} was first measured by applying a small-amplitude
cyclic load. The bedding error generated by the interfaces
between the specimen and the loading upper/lower plates
has been found to be minimal for the cyclic test.

Shear Modulus and Hysteretic Damping

In Fig.12 the small strain shear modulus, G_0,
measured for all tested samples under the in-situ effective
overburden are plotted against the void ratio, e.   the
increase in the modulus with decreasing void ratio can be
obviously seen, but the trend is quite different from one
gravel to others.   The KJ-site specimens give the highest
modulus while the A-site specimens do the lowest.   As
indicated with the close marks, the A-site specimens
obtained by the conventional non-freezing sampling method
show still lower modulus implying some larger disturbing
effect during the conventional sampling procedure.   In
the figure the plots are approximated by the well-used
formula   (Richart et al. 1970); $G_0=[(2.17-e)^2/(1+e)]K$,
leading to widely different values for K.   This difference
in K may be attributed mainly to the difference in
geological history of these gravel layers.

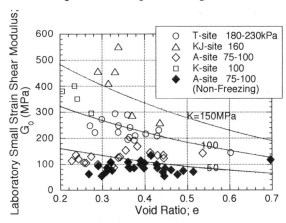

Fig.12 Laboratory Small-Strain Shear Modulus(G_0)
       vs Void Ratio(e)

The laboratory small strain modulus, G_0, are then
compared in Fig.13 with the shear modulus calculated from
the in-situ shear wave velocity and the saturated density
of the gravel.   In T-site where the in-situ shear-wave
velocity was measured in a finer pitch by the suspension
method, the comparison are made plot by plot, while, in
other sites where the down-hole method was employed to
measure the average of the gravel layer, a representative
single value is taken in the vertical axis for each site.
Despite the lack of local wave velocities in most sites,
this figure clearly indicates the modulus based on in-situ
wave velocity is apparently larger than the laboratory
modulus, and the gap between them increases for larger
modulus values.   This gap seems unlikely to be filled
solely by the errors involved in laboratory test measurement.
However , compared with conventional samplings, it obviously
provides highly intact sample as indicated by the comparison
of the two kind of test results in A-site shown in Fig.13.

Fig.14 shows the variations in the small-strain
modulus , G_0, in the full-logarithmic graph due to the
increase in the confining effective stress measured for
thirteen intact specimens taken from three sites.   The
variations are approximated by the following equation;
$G_0=(\sigma_c'/p_0)^n K$, as shown by the straight lines in the same
figure.   the power in the equation , n, as evaluated and
listed in the figure takes 0.56 to 0.93 much higher than
that normally measured for clean sands (0.4 to 0.5) or
reconstituted gravels (0.55 to 0.60) (Kokusho 1982).   It
is further noted that the power, though scattered, clearly
differs from site to site; T-site is the highest, seemingly
reflecting difference in some kind of soil structures.

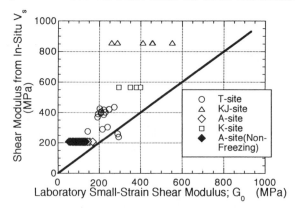

Fig.13 Comparison of Small-Strain Shear Modulus
       between In-Situ and Laboratory

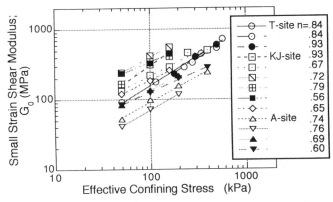

Fig.14 Relationship between Small-Strain Modulus(G_0)
       and Confining Stress(σ_c')

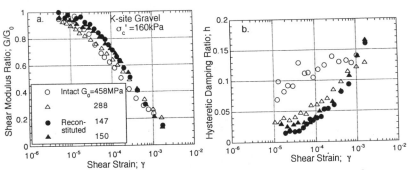

Fig.15 Shear Modulus Ratio(a) and Damping Ratio(b)
       for Intact and Reconstituted Gravel

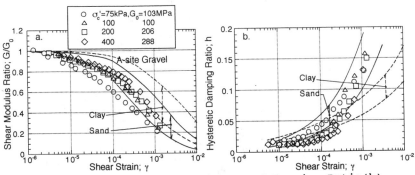

Fig.16 Shear Modulus Ratio(a) and Damping Ratio(b)
       of Gravel for Different Confining Stress

In Fig.15a the strain-dependent modulus degradation
curves measured for K-site intact specimens are compared
with those for reconstituted specimens having almost the
same void ratio on the semi-logarithmic gragh.   The
differences are obvious between the two kind of specimens
not only for the small-strain modulus, G_0, but also for
the degradation of the modulus ratio; the intact specimen
exhibits greater degradation for smaller strain level.
Fig.15b shows the hysteretic damping ratio measured in the
same series of tests.   The damping ratio of the intact
sample is obviously larger than that of the reconstituted
one.

Figs.16a and b show the shear modulus ratio and the
damping ratio respectively plotted against the shear strain
amplitude for four intact specimens from A-site under four
different confining stresses.   The variations of the
modulus ratio are evidently influenced by the confining
stress in the same way as sands shown by previous researchers
(e.g. Kokusho 1983).   In the same graph, the degradation
curves for clean sands and cohesive soils are illustrated,
indicating that the degradation is more pronounced for
gravels than other soils for smaller strain levels.

Undrained Cyclic Strength

In Fig.17 the undrained cyclic strength (the stress
ratio,$\sigma_d/2\sigma_c{}'$(σ_d=dynamic   stress,$\sigma_c{}'$= effective confining
stress) for attaining 2% double amplitude strain in a
certain number of cycles, N_c) is plotted against the
number of loading cycles, N_c, on the semi-logarithmic
graph for the A-site specimens both by in-situ freezing
and conventional non-freezing sampling.   Though the
strengths for the freezing sampling specimens are widely
scattered, they are obviously larger than those for
conventionally sampled specimens, demonstrating the
effectiveness of the in-situ freezing sampling method.

The stress ratios for all in-situ freezing sampling
specimens from four sites are plotted against N_c on the
full-logarithmic graph in Fig.18.   They take the value
of 0.3 to more than 1.0 for N_c=20.   Despite the wide
scatters in the group of specimens from the same site, the
KJ or K-site specimens yield obviously higher strength
than the T or A-site specimens.   Assuming that the plots
between the stress ratio versus the number of cycles on
the logarithmic graph may be approximated by a straight
line as proposed by Annaki and Lee (1977), its slope has
been evaluated as -0.12 to -0.22 for three specimen groups
as shown in the figure.   Based on these values the slope
roughly postulated as -0.2 has been used in the later

data interpretation for estimating the stress ratios corresponding to N_c=20 for individual specimens in order to eliminate individual soil conditions in each group of specimens.

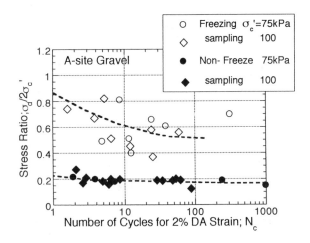

Fig.17 Comparison of Undrained Cyclic Strengths for Freezing and Non-Freezing Sampling

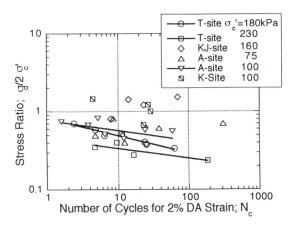

Fig.18 Undrained Cyclic Strength vs Number of Cycles for 2% DA-Strain for All Specimens

The stress ratios, $R(N_c=20)$, thus estimated are shown in
Fig.19 against the relative density, D_r, for each specimen
from two sites for which the data on D_r is available. D_r
here has been determined with the method as described
before. It is remarkable that the estimated stress ratio
seems to be rather consistent with the relative density
and that the reconstituted specimens of KJ-site exhibits
only a fraction of strength compared to the intact specimens
for the same relative density. The difference between
the intact specimens and the reconstituted ones tend to
enlarge with increasing relative density in a similar
manner as indicated for dense sands by Kokusho et al.(1985).

Fig.20 shows the relationship between the estimated
stress ratio, R_{20}, and the small-strain modulus, G_0, measured
for each specimen from the four sites. One can readily
find some kind of close correlations between the two
variables, apparently indicating a possibility to estimate
the in-situ undrained cyclic strength from the in-situ
shear modulus, hence the in-situ shear-wave velocity,
despite the fact that both variables may somewhat differ
from in-situ values. However one correlation seems valid
only for one site and not for other sites, implying that
one needs to establish a site-dependent correlation to
estimate the ubdrained cyclic strength from the shear wave
velocity. Fig.21 further evidences this point. In this
figure the stress ratio directly read off from Fig.18 for
$N_c=20$ for each gravel layer in the four sites is plotted
against the in-situ shear wave velocity measured by the
down-hole method. the relationship doesn't seem universally
applicable to estimate the undrained cyclic strength from
the shear wave velocity.

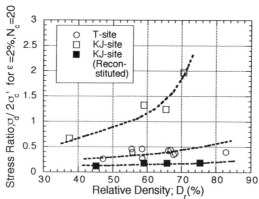

Fig.19 Estimated Undrained Cyclic Strength($\varepsilon=2\%$, $N_c=20$)
      vs Relative Density

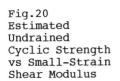

Fig.20
Estimated
Undrained
Cyclic Strength
vs Small-Strain
Shear Modulus

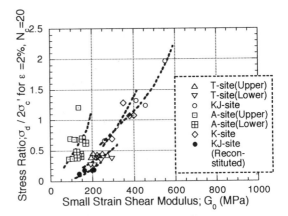

Fig.21
Stress Ratio
vs In-Situ
Shear-Wave
Velocity

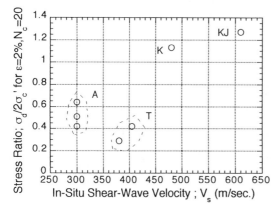

Fig.22
Stress Ratio
vs Penetration
Test Blow Counts

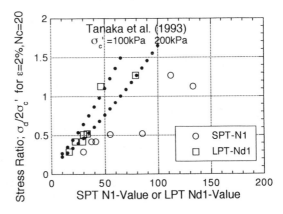

Finally the undrained cyclic strength is plotted against the SPT N_1-value (the normalized blow counts for =100kpa) and the LPT Nd_1-value measured in the same gravel layers, as shown in Fig.22. Quite obviously the stress ratio is more closely related to the penetration test blow-counts than to the shear wave velocity, and the LPT blow-counts are more consistent than SPT blow counts presumably because the LPT blow-counts more reliably reflect the soil structure with less effect of gravel size.

Therefore, the LPT blow-counts are the most promising index to roughly estimate the in-situ undrained cyclic strength of gravel layers. However, an important problem to be addressed for that is the fact that the stress ratio tends to decrease with increasing confining stress . Tanaka et al.(1993) based on these data and other researcher's data made a comprehensive study to take account of the effect of this effect and proposed the following equation;

$$\sigma_d \ /2\sigma_c' = \ 0.15 + 0.0059(Nd_1 p_1/\sigma_c')^{1.3}$$, where p_1 = the unit

pressure(100kpa). The curves for σ_c'=100kpa and 200kpa are shown in the same figure, demonstrating the applicability of this empirical formula.

Lastly it should be commented that the in-situ freezing sampling, too, may not be a perfect method to recover an ideally intact sample. Namely Kataoka et al.(1989) showed by laboratory tests using reconstituted gravels that the undrained cyclic strength will be lowered by maximum 20% due to the freeze/thaw process if the gravels are over-consolidated. Therefore, considering that Pleistocene gravels normally have the cementation effect similar to the over-consolidation effect, it may well be expected the undrained cyclic strength evaluated in this research still possesses some amount of safety margin.

Conclusions

In order to clarify in-situ dynamic properties of gravelly soils, it is indispensable to sample soils as intact as possible. For this purpose the in-situ freezing sampling method has been developed and applied to Pleistocene gravel layers in four different sites, revealing the following major facts;

(1) The in-situ void ratios of Pleistocene gravels are mostly around 0.25 to 0.45 and tend to be smaller with the increase in the uniformity coefficient or the gravel content.
(2) The effectiveness of the in-situ freezing sampling has been fully demonstrated in measuring the soil modulus and

damping as well as the undrained cyclic strength by comparing the test results with those of conventionally sampled specimens or of reconstituted specimens.
(3) There exists no universal correlation between the shear modulus (or the shear wave velocity) and the void ratio. The correlations are quite different from site to site probably due to the difference in soil structure or geological history.
(4) The undrained cyclic strength can not be universally evaluated from the in-situ shear wave velocity or the relative density though for each site there possibly exists an unique relationship between them.
(5) It has been found that, like sands, the undrained cyclic strength of gravel layers can be estimated by the in-situ penetration test results, preferably by the LPT blow-counts.

Acknowledgment

The authors are grateful to Dr. K. Sakai and other members of Kisojiban Consultants Co. Ltd. for their great contributions in developing and implementing the in-situ freezing sampling technique.

References

1. Andrus,R.D. and Youd,T.L.,"Penetration Tests in Liquefiable Gravels," Proc. 12th Int. Conf. on SMFE Vol.1, pp. 679-682, 1989.
2. Annaki, M. and Lee, K.L., "Equivalent Uniform Cycle Concept for Soil Dynamics," Proc.ASCE, Vol.103, GT6, pp549-564, 1977.
3. Ikemi,M., Kudo, K. and Kokusho, T. ,"On Relative Density of Gravels," Proc. 19th JSSMFE, (in Japanese), 1984.
4. Ishihara,K., "Fundamentals of Soil Dynamics," Kashima Publishers Inc., (in Japanese),1976.
5. Ishihara,K, Kokusho, T. and Silver,M.L., "Recent Developments in Evaluating Liquefaction Characteristics of Local Soils" State-of-the-Art Report Pro. 12th Int. Conf. on SMFE, pp2719-2734, 1989
6. Kataoka,T., Yoshida,Y., Okamoto,T. and Kokusho,K., "Applicability and Development of In-Situ Freezing Sampling Method to Sandy and Gravelly Soils" CRIEPI Report No.U88073, (in Japanese), 1989.
7. Kokusho,T., "Cyclic Triaxial Test of Dynamic Soil Properties for Wide Strain Range," Soils and Foundation, Vol.20, No.2, pp.45-60, 1980.

8. Kokusho,T., "Dynamic Soil Properties of and Nonlinear
   Seismic Response of Ground" CRIEPI General ReportNo.301
   (in Japanese), 1982.
9. Kokusho,T., Yoshida,Y. and Nagasaki,K.,"Liquefaction
   Strength Evaluation of Dense Sand Layer," Proc. 11th
   Int. Conf. on SMFE, pp.1897-1900, 1985.
10.Kokusho,T., ''In-Situ Dynamic Properties and Their
   Evaluations," Proc. 8th Asian Regional Conf. on SMFE,
   Vol.2, pp215-240, 1987.
11.Kokusho,T., Nishi,K., et al., "Study on Quaternary
   Ground Siting of Nuclear Power Plant, Part-I:
   Geological/Geotechnical Investigation Methods and
   Seismic Stability Evaluation Methods of Foundation
   Ground," CRIEPI General Report No.U19, (in Japanese),
   1991.
12.Kokusho,T., Tanaka,Y. and Kawai,T., "Liquefaction Case
   Study of Volcanic Debris Flow Layer during 1993
   Hokkaido-Nanseioki-Earthquake," Proc. 9th Japan
   Earthquake Eng. Symp. (to be published), 1994.
13.Richart,F.E., Hall,J.R. and Woods,R.D., "Vibrations of
   Soils and Foundations," Prentice-Hall Inc., 1970
14.Tanaka,Y., Kudo,K., Yoshida,Y., Kataoka,T. and
   Kokusho,T., "A study on the Mechanical Properties of
   Sandy Gravel-Mechanical Properties of Undisturbed
   Sample and Its Simplified Evaluation," CRIEPI Report
   No.U88021, (in Japanese), 1988.
15.Tanaka,Y., Kudo,K., Yoshida,Y., Nishi,K. et al., "On
   the Applicability of Various Sampling Methods to the
   Gravelly Ground," CRIEPI Report No.U90046,(in
   Japanese), 1990.
16.Tanaka,Y., Kokusho,T, Yoshida,Y. and Kudo,K., "A Method for
   Evaluating Membrane Compliance and System Compliance in
   Undrained Cyclic Shear Tests," Soils and Foundations Vol.31,
   No.3, pp.30-42, 1991.
17.Tanaka,Y., Kudo,K., Yoshida,Y. and Kokusho,T.,
   "Evaluation Method of Dynamic Strength and Residual
   Settlement for Gravelly Soils," CRIEPI Report
   No.U90063, (in Japanese), 1991
18.Tanaka,Y., Kudo,K., Yoshida,Y. and Kokusho,T.,
   "Undrained Cyclic Strength of Gravelly Soil and Its
   Evaluation by Penetration Resistance and Shear
   Modulus," Soils and Foundations, Vol.32, No.4, pp.128-
   142, 1992.
19. Yoshida,Y. and Kokusho,T., "Empirical Formulas of SPT
   Blow-Counts for Gravel Soils," Proc. ISOPT-1 for
   Penetration Testing, pp.381-387, 1988.

DYNAMIC PROPERTIES OF GRAVELS SAMPLED BY GROUND FREEZING

Shigeru Goto[1], Shin'ya Nishio[2]
and Yoshiaki Yoshimi[3], F. ASCE

Abstract

This paper describes two methods to obtain high-quality undisturbed samples of gravel by freezing it in situ. The quality of the undisturbed samples obtained by these methods is judged good because the requirements previously established for obtaining high-quality undisturbed samples of clean sand are satisfied and because the shear wave velocities measured in the laboratory agree with those measured in the field. Two kinds of gravels, one alluvial and the other Pleistocene, were sampled and tested in the laboratory to determine their dynamic properties. The test results on the undisturbed samples are compared with the results on samples reconstituted to the same density. Shear modulus at very small shear strain, G_0, of the undisturbed samples is greater than that of the reconstituted samples, but the relationship between normalized shear modulus G/G_0 and shear strain of the undisturbed samples is almost the same as the reconstituted samples. The liquefaction resistance of the undisturbed samples is considerably higher than that of the reconstituted samples. For the Pleistocene gravel the difference in liquefaction resistance between the undisturbed samples and reconstituted samples is much greater than that for the alluvial gravel. The shear modulus and liquefaction resistance of the Pleistocene gravel are significantly greater than those of the alluvial gravel, reflecting the greater penetration resistance of the former.

Introduction

Extensive studies have been made to show that high-quality undisturbed samples of clean sands can be obtained by freezing the ground in situ if the following requirements are met: (1) Fines content is low, (2) confining pressure is moderately high,(3) freezing progresses without impeding drainage at the freezing front, and (4) the possible zone of disturbance due to drilling and the insertion of the freezing pipe is avoided (Yoshimi et al., 1978; 1989). The sample thus obtained is said to be of high-quality because it retains the in situ values of density, elastic shear modulus and undrained cyclic strength (liquefaction resistance). The claim to the high-quality concerning liquefaction resistance is supported by laboratory tests showing that the

---

[1] Senior Research Engineer, Institute of Technology, Shimizu Corporation, 4-17, Etchujima 3-chome, Koto-ku, Tokyo, Japan; [2] Research Engineer, Ditto; [3] Senior Advisor, Shimizu Corporation, Tokyo, Japan; Professor Emeritus, Tokyo Institute Technology, Tokyo, Japan

liquefaction resistance is not affected by a history of unidirectional freezing and subsequent thawing (Singh et al., 1982; Goto, 1993), and by the fact that the correlation between the liquefaction resistance and corrected Standard Penetration Test (SPT) N-values is compatible with a correlation based on field performance data during real earthquakes (Yoshimi et al., 1989), and that the samples retain their in situ shear wave velocity which is related to liquefaction resistance (Tokimatsu et al., 1986).

Although similar studies on gravel are still limited, one can deduce that the requirements for insuring high quality of its in situ frozen samples can be satisfied with gravel if the sampling and testing procedures are properly modified to take the larger grain size into account.

The objectives of this paper are to describe two methods of sampling for clean gravel by in situ freezing and to present the results of undrained cyclic tests on the in situ frozen samples and the samples reconstituted to the same density.

Description of the Sampling Sites and Field Test Results

Undisturbed samples of gravel were obtained at two sites: Site A near the Tone River in Saitama Prefecture, Japan, and Site B in Chiba Prefecture, Japan. The sampling depths at Site A were between 5 m and 15 m (Fig. 1), and those at Site B between 6 m and 8 m (Fig. 2). The gravel at Site A is an alluvial deposit while the gravel at Site B is a Pleistocene deposit. As shown in Fig. 2 the upper part of the sampled gravel was above the ground water table, and water had to be injected during the ground freezing process (Goto et al., 1987).

The field tests included the Standard Penetration Test (SPT), Large Penetration Test (LPT) and geophysical survey. The specifications for the LPT are compared in Table 1 with those for the SPT. Both types of penetration tests were conducted by a free fall method of hammer release using a trigger mechanism locally called "tombi." Note that the SPT blow counts more than 50 were extrapolated from the penetrations at 50 blows that were less than 30 cm, thus should be considered less reliable than the LPT blow counts that were not extrapolated. The shear wave velocities were determined by the down-hole method.

TABLE 1. Specifications for Penetration Tests (Suzuki et al., 1993)

			SPT	LPT
Hammer	Mass,	kg	63.5	100
	Fall height,	mm	750	1500
Rod	OD,	mm	40.5	60
Sampler	OD,	mm	51	73
	ID,	mm	35	50

Methods of Ground Freezing

Fig. 3 shows the method of ground freezing and subsequent coring at Site A where liquid nitrogen was circulated in a single freezing pipe.This procedure for ground freezing insures free drainage at the freezing front, and is similar to other applications to sand deposits as described by Yoshimi et al. (1984, 1989), except for a

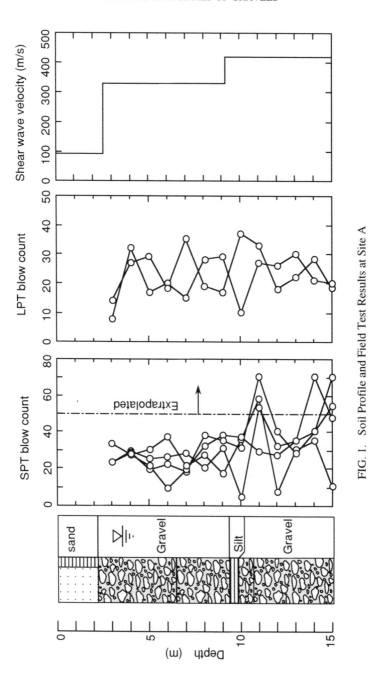

FIG. 1. Soil Profile and Field Test Results at Site A

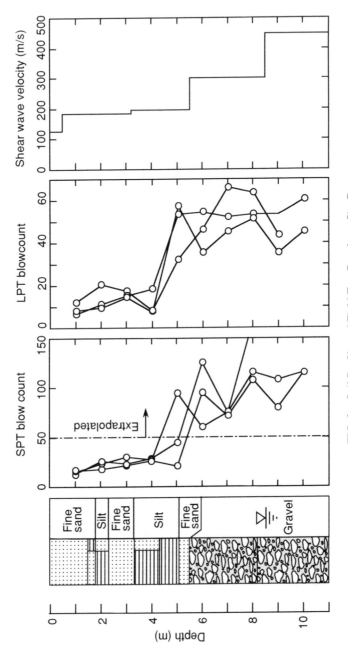

FIG. 2. Soil Profile and Field Test Results at Site B

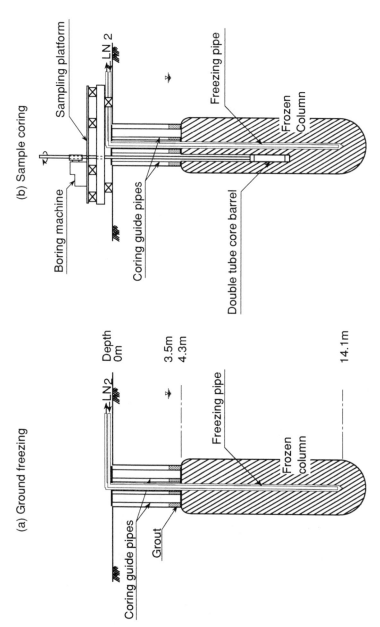

(b) Sample coring

(a) Ground freezing

FIG.3. Method of Sampling by In Situ Freezing at Site A

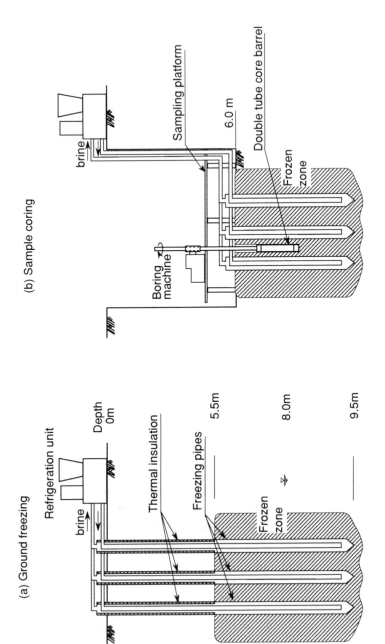

FIG. 4. Method of Sampling by In Situ Freezing at Site B

larger radius required for gravel for two reasons: the possible zone of disturbance due to drilling is likely to be larger, and much larger specimens are required. It took eight days to freeze the ground to a radius of a little over 1 m.

At Site B a different method as shown in Fig. 4 was used where calcium chloride brine (-30°C) was circulated in five freezing pipes. It took about four weeks to freeze an adequate amount of the gravel.

Fig. 5 shows the layout for monitoring the temperatures in the ground. With this system the ground temperatures measured at a predetermined interval, say every 30 minutes, are automatically recorded and displayed on the CRT as time histories. In addition, the microcomputer enables one to predict the temperatures at coring locations as shown in Fig. 6 (Goto et al., 1986).

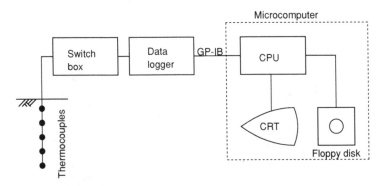

FIG. 5. Temperature Monitoring System for Ground Freezing

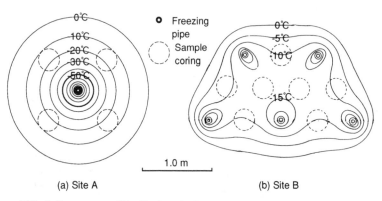

FIG. 6. Temperature Distributions in Frozen Ground at the Time of Coring

Coring from the Frozen Gravel

At both sites the coring was done with a 300-mm double tube core barrel with diamond bits using -15°C drilling fluid. Cores about 1 m long were obtained at four or seven locations at least 300 mm away from the freezing pipe to avoid the zone of possible disturbance as shown in Fig. 6. The acquisition of several specimens from the same depth was a definite advantage for establishing well-defined shear stress vs. number of cycles curves (e.g. Fig. 13) to determine liquefaction resistance. Unlike sand specimens, typically 75 mm in diameter, that are machined in the laboratory from a large block of frozen sand, the gravel specimens were cut into their exact size on site. The cylindrical surface of the core was very smooth like a concrete core because the diamond bits cut cleanly through individual gravel particles. The smooth surface is highly desirable for minimizing membrane penetration effects during undrained cyclic tests. Each core was cut to a length of 600 mm with a diamond blade circular saw that also produced very smooth surfaces at the ends. This is desirable for insuring good fit with the pedestal and the top cap of the triaxial apparatus for minimizing errors in axial strain measurements.

(a) Side view of core at Site A   (b) Sawed end of core at Site B

PHOTO. 1. Core of Frozen Gravel

The contours in Fig. 6 show theoretical temperature distributions in the horizontal plane. The calculated temperatures agreed with the measured values within 0.83°C (Goto, et al., 1986). Note that the temperature gradient within the cores is steeper at Site A. In general multiple freezing pipes may cause an undesirable heave when a part of the soil trapped within a frozen zone undergoes delayed freezing for lack of free drainage (Osterberg and Varaksin, 1973). In the case of Site B, however, such a problem was not likely to have occurred because the gravel was very permeable and partially saturated in the upper half.

Each frozen core was covered with 2-mm rubber membrane and placed in a rigid steel container which was lined with 50-mm thick heat insulating material to prevent the melting and sublimation of the pore ice. The core was kept frozen during transport and storage at about -20°C until it was ready to be placed in the triaxial apparatus.

Physical Properties of Gravel Samples

Table 2 and Fig. 7 show the physical properties of the gravels sampled. The gravel at Site A is considerably coarser and better graded than the gravel at Site B. Both gravels have nearly the same void ratios and very low fines contents.

TABLE 2. Physical Properties of Gravels

	Maximum grain size (mm)	Mean grain size (mm)	Cofficient of uniformity	Void ratio	Gravel content (%)	Fines content (%)	Geological age
Site-A	105	10.9	39	0.34	66.4	0.3	Alluvial
Site-B	94	1.5	8.2	0.39	54.7	1.9	Pleistocene

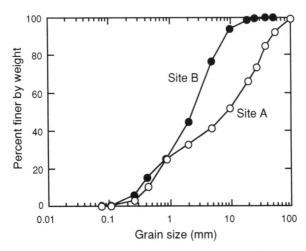

FIG. 7. Average Grain Size Distribution Curves of the Gravels Sampled

Procedures for Thawing and Saturating Specimens

All dynamic tests were run with a large triaxial apparatus shown in Fig. 8 that can accommodate a specimen 300 mm in diameter and 600 mm in height (Goto et al., 1992). The instruments consist of a load cell, a pair of displacement transducers (Bison gauges) and a gap sensor that are placed within the cell, and an LVDT that is mounted outside the cell.

A frozen specimen in rubber membrane, 300 mm in diameter and 600 mm in height, was placed in the triaxial cell, and subjected to an isotropic pressure of 29 kPa for thawing by circulating lukewarm water of 40°C around the specimen. After thawing that took about 2 hours, the specimen was saturated by the usual combination of CO_2 gas, deaired water and back pressure until the pore pressure coefficient B-value exceeded 0.95.

Measurement of Shear Wave Velocity in the Laboratory

The shear wave velocities of the gravel specimens at very small strain level

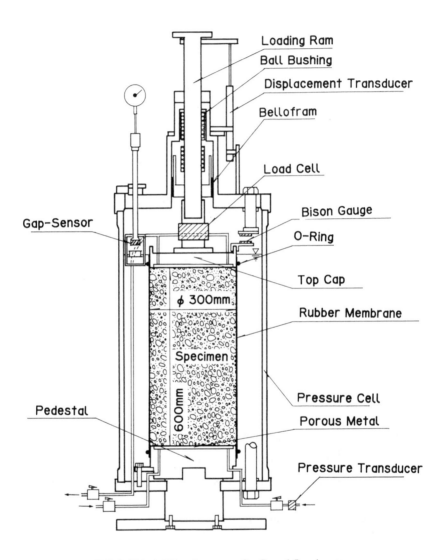

FIG. 8. Triaxial Test Apparatus for Gravel Specimens

were determined in the triaxial cell by lightly tapping the side of the top cap and measuring the arrival times of the shear waves at four accelerometers fixed on the surface of the specimen as shown in Fig. 9. This method has two advantages: (1) It simulates the shear wave propagation in the down-hole method in the field, and (2) the time intervals are measured within the specimen and thus unaffected by possible lack of perfect contact between the top cap and the specimen.

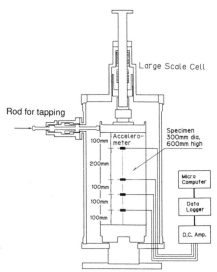

FIG. 9. Layout for Shear Wave Velocity Measurements in the Laboratory

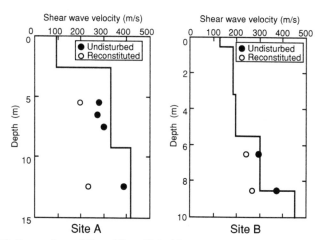

FIG. 10. Comparison of Shear Wave Velocities Measured in the Field and in the Laboratory

The data points in Fig. 10 show the results of tests on the specimens isotropically consolidated to the in situ overburden pressures for both the undisturbed and reconstituted samples. This implies that the horizontal effective stress in the field was assumed equal to the vertical effective stress. The lines in Fig. 10 show the field test results reproduced from Figs. 1 and 2 for comparison. Although the data are too few to warrant a definite conclusion, the shear wave velocities of the undisturbed samples are reasonably close to the field values, whereas those of the reconstituted samples are smaller than the field values, particularly for the gravel at Site A.

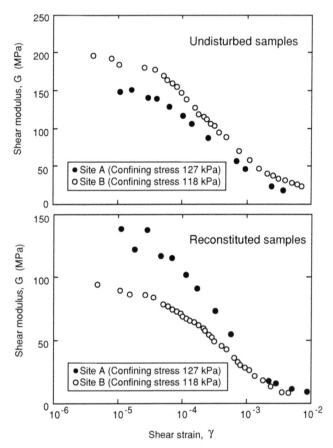

FIG. 11. Relationship Between Shear Modulus and Shear Strain Amplitude

Effect of Shear Strain Amplitude on Shear Modulus

A series of cyclic loading tests were run to study the effect of shear strain amplitude on shear modulus. Accurate measurements of deviator stresses and axial strains for low amplitudes were possible because the transducers were placed inside the pressure chamber as shown in Fig. 8. The tests were run in stages by gradually increasing the amplitude of deviator stress. Ten cycles of completely reversed deviator stresses were applied under undrained conditions, and the data of the 5th cycle was used to compute shear strain by assuming a Poisson's ratio of 0.5. Between successive loadings the specimen was drained to restore the initial effective confining stress.

Fig. 11 compares the shear modulus vs. shear strain curves of the gravels at Sites A and B for the undisturbed sample and the reconstituted sample. Such comparison is considered meaningful because the confining stresses for the two gravels are nearly equal. The greater stiffness of the undisturbed samples of the Site B

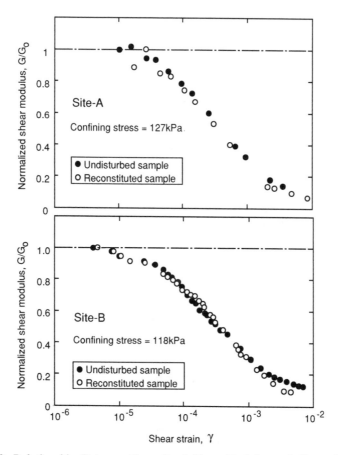

FIG. 12. Relationship Between Normalized Shear Modulus and Shear Strain Amplitude

gravel is consistent with their higher penetration resistance as shown in Figs. 1 and 2. Concerning the reconstituted samples, on the contrary, the Site A gravel is stiffer than the Site B gravel.

Fig. 12 compares the undisturbed and reconstituted samples concerning normalized shear modulus, G/G_0. No significant difference is visible for the whole range of shear strains. This kind of agreement between the undisturbed and reconstituted samples was reported by Tokimatsu and Hosaka (1986) for a clean, dense sand, although the shape of their curve was somewhat different. If a common G/G_0 vs. γ curve exists in general for undisturbed and reconstituted samples having the same density, one can obtain the G vs. γ curve for the in situ soil by using the in situ G_0 that can be computed from the shear wave velocity measured in the field from the well known relationship, $G_0 = \rho V_s^2$, in which ρ is soil density. For this method to

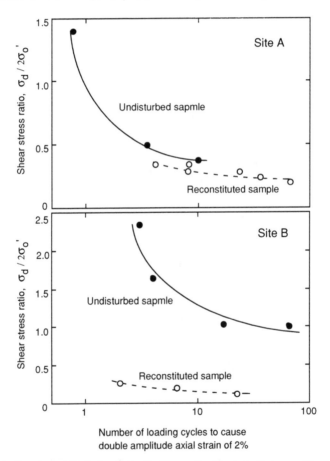

FIG. 13. Results of Undrained Cyclic Tests: Comparison Between Undisturbed Sample and Reconstituted Sample

be practical, however, one should be able to determine the in situ density by a simpler method than sampling by in situ freezing, e.g. by nuclear density logging.

Liquefaction Tests

Fig. 13 shows the results of undrained cyclic triaxial tests on the undisturbed and reconstituted samples. Because the cylindrical surfaces of the reconstituted specimens were not as smooth as the frozen specimens, there remained a possibility of overestimating the undrained strength due to membrane penetration effects. Thus the ordinates of Fig. 13 for the reconstituted samples may represent an upper limit.

For both gravels the undisturbed samples are stronger than the reconstituted samples, although the difference is much more pronounced for the gravel at Site B.

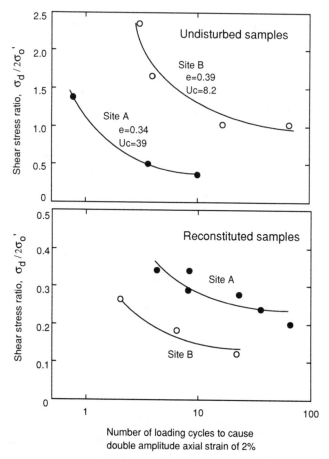

FIG. 14. Results of Undrained Cyclic Tests: Comparison Between Site A and Site B

Because the undisturbed and reconstituted specimens were of the same density, the difference in their undrained strengths must be attributed to some difference in soil fabric which may be related to geological age. Obviously one will greatly underestimate the liquefaction resistance of the Site B gravel by relying on the test results on the reconstituted samples.

The test data of Fig. 13 are replotted in Fig. 14 to enable direct comparison between the two gravels. The relative positions of the curves are similar to Fig. 11 for shear modulus, although the differences are greater in the case of liquefaction resistance, particularly for the undisturbed samples. The greater strength of the undisturbed samples of the Site B gravel is consistent with their higher penetration resistance as shown in Figs. 1 and 2.

Conclusions

The following conclusions may be drawn on the basis of shear wave velocities measured in the field and in the laboratory, and on the results of undrained cyclic shear tests on in situ frozen samples and reconstituted samples of two gravels in Japan.

(1) The quality of the undisturbed samples of the gravels obtained by the in situ freezing method is judged good because the requirements previously established for obtaining high-quality undisturbed samples of clean sand are satisfied, and because the shear wave velocities measured in the laboratory agree with those measured in the field. Thus the in situ frozen samples of gravels reported in this paper retain their density, shear modulus and undrained cyclic strength in situ.
(2) The samples that were reconstituted to the same density as the in situ frozen samples had lower shear modulus and undrained cyclic strength than the in situ frozen samples, the difference being probably dependent on geological age.
(3) The normalized shear modulus vs. shear strain curves for the reconstituted samples were nearly the same as those for the in situ frozen samples.

Appendix, References

Goto, S. (1993). "Influence of a freeze and thaw cycle on liquefaction resistance of sandy soil." *Soils and Foundations*, Tokyo, Japan, 33(4), 148-158.
Goto, S., Nishio, S., Shamoto, Y., Akagawa, B. and Tamaoki, K. (1986). "Undisturbed sampling by in-situ freezing and dynamic properties of diluvial gravel (in Japanese)." *Technical Research Report of Shimizu Construction Co., Ltd.*, 44, 13-23.
Goto, S., Shamoto, Y. and Tamaoki, K. (1987). "Dynamic properties of undisturbed gravel sample by in-situ frozen [*sic*]." *Proc. 8th Regional Conf. on Soil Mech. and Found. Eng.*, Kyoto, Japan, 1, 233-236.
Goto, S., Suzuki, Y., Nishio, S. and Oh-oka, H. (1992). "Mechanical properties of undisturbed Tone-River gravel obtained by in-situ freezing method." *Soils and Foundations*, Tokyo, Japan, 32(3), 15-25.
Osterberg, J. O. and Varaksin, S. (1973). "Determination of relative density of sand below ground water table." *Special Technical Publication 523*, Am. Soc. Testing and Materials, 364-378.
Suzuki, Y., Goto. S., Hatanaka, M. and Tokimatsu, K. (1993). "Correlation between strengths and penetration resistances for gravelly soils." *Soils and Foundations*, Tokyo, Japan, 33(1), 92-101.

Tokuimatsu, K. and Hosaka, Y. (1986). "Effects of sample disturbance on dynamic properties of sand." *Soils and Foundations*, Tokyo, Japan, 26(1), 53-64.

Tokimatsu, K., Yamazaki, K. and Yoshimi, Y. (1986). "Soil liquefaction evaluations by elastic shear moduli." *Soils and Foundations*, Tokyo, Japan, 26(1), 25-35.

Yoshimi, Y., Tokimatsu, K., Kaneko, O. and Makihara, Y. (1984). "Undrained cyclic shear strength of a dense Niigata sand." *Soils and Foundations*, Tokyo, Japan, 24(4), 131-145.

Yoshimi, Y., Tokimatsu, K. and Hosaka, Y. (1989). "Evaluation of liquefaction resistance of clean sands based on high-quality undisturbed samples." *Soils and Foundations*, Tokyo, Japan, 29(1), 93-104.

CYCLIC BEHAVIOR OF GRAVELLY SOIL

Mark D. Evans[1], Member, ASCE
and Shengping Zhou[2]

ABSTRACT

Gravelly soils have been shown to be liquefiable by a number of investigators. Undrained cyclic triaxial testing may reliably predict cyclic behavior if membrane compliance is eliminated. Loose, uniformly-graded gravel is shown to have similar cyclic strength as sand in this paper once membrane compliance is mitigated. A correction for membrane compliance is also presented.

When gravelly-sand is encountered, undrained, cyclic triaxial tests may be performed on large-scale specimens or on smaller specimens where the oversized particles have been removed. However, the dynamic properties of gravelly sand are significantly affected by the gravel particles, even if the gravel is floating in a sandy matrix. This paper shows the significant effect of gravel inclusions on the dynamic properties of sand-gravel composite specimens. Results of tests on sand-gravel composites with 0%, 20%, 40%, and 60% gravel content show an increase in cyclic strength with increased gravel content, even though the relative density of the composite is constant. Differences in stress--strain behavior, axial strain, and pore pressure ratio are illustrated for soils with different gravel contents. Neither the overall composite relative density nor the matrix soil relative density are reliable indicators of dynamic properties of gravelly soil at different gravel contents. The composite specimen behaves like the matrix soil at a higher relative density. A methodology is described for estimating the cyclic loading resistance of sand-gravel composites based on its normalized void ratio function.

---

[1] Asst. Prof., Civil Eng. Dept., Northeastern Univ., 420 SN, Boston, MA 02115.
[2] Graduate Res. Asst., Civil Eng. Dept., Northeastern Univ., Boston, MA.

INTRODUCTION

In recent years, the liquefaction behavior of gravelly soil has been investigated in the laboratory (Nicholson et al., 1993; Evans, 1993; Evans et al., 1992; Seed et al., 1989; Hynes, 1988; Banerjee et al., 1979; and Wong et al., 1974). Gravelly soils have been shown to be liquefiable in situ if drainage is impeded; and reasonable cyclic strength values may be estimated in the laboratory if membrane compliance is considered. However, field evidence has shown that most liquefied gravelly soils are sand-gravel composites (Evans and Harder, 1993; Harder and Seed, 1986). There have been few studies concerned with the liquefaction behavior of sand-gravel composites. In order to quantify the effect of gravel content on the liquefaction resistance of gravel and sand-gravel composites, a series of undrained cyclic triaxial tests were performed in this study on gravel, sluiced gravel, and sand-gravel composite specimens with gravel contents of 0%, 20%, 40%, and 60%. These data are compared with data for sand specimens.

LABORATORY CYCLIC STRENGTH OF GRAVEL

Introduction

The cyclic strength of gravel has been studied by the authors and by other investigators. Such analyses may include performing undrained, cyclic triaxial tests to determine liquefaction resistance and dynamic material properties. However, membrane penetration and compliance effects must be considered when testing such materials. Membrane compliance may result in pore fluid redistribution, soil densification, and increased liquefaction resistance in the undrained triaxial test, thus, making it difficult to properly evaluated the performance of the prototype material in situ. Several investigators have proposed methods accounting for membrane compliance. Nicholson et al. (1993a, 1993b) proposed a method for changing the volume of the specimen during cyclic testing to account for membrane compliance. Evans et al. (1992) proposed sluicing uniformly-graded gravel specimens with sand to minimize membrane compliance prior to cyclic testing. The latter method will be discussed briefly below.

Density Changes Due To Membrane Compliance

A reduction in effective confining pressure during undrained, cyclic testing causes the membrane to rebound from penetration sites, pore fluid to migrate to the specimen periphery, and the skeletal grain structure to contract to balance the fluid volume. Thus, specimen density at the end of undrained cyclic loading may be considerably higher than the density at the start, even though the test is undrained. Changes in specimen density are caused by the compliant nature of the rubber confining membrane during undrained loading. The test system remains undrained and the volume of the system is constant throughout the test. However, the specimen volume changes during undrained testing as pore fluid is redistributed between internal and peripheral void spaces due to membrane compliance.

Density changes in undrained tests caused by membrane compliance may be computed from membrane penetration volume changes measured during drained hydrostatic rebound. Volumetric strain values may then be converted to corresponding increases in relative density. Fig. 1 shows the increase in relative density of gravel specimens versus residual pore pressure ratios developed in the test. Once the residual pore pressure ratio is determined at the end of a test, one can determine the increase in relative density caused by membrane compliance. Note that Fig. 1 is specific to the material and membrane properties tested and are not universally applicable. These data may be determined quickly and easily at the start of each new testing program. However, for the purposes of this paper, it may be seen that the relative density of gravel specimens may increase by up to 20 percentage points or more for some test conditions. A 30% relative density gravel specimen developing 80% residual pore pressure at failure, for example, would increase in relative density by almost 19 percentage points to 49%. Thus, although the intent was to test a specimen with a relative density of about 30%, the relative density gradually increased to 49% due to membrane compliance. The resulting value of cyclic loading resistance is, therefore, erroneously high and unconservative and does not accurately represent the material property in situ.

Cyclic Loading Resistance of Sluiced and Unsluiced Gravel

Artificial increases in specimen density, caused by membrane compliance, increase the cyclic loading resistance of the soil. In an effort to reduce membrane compliance, Evans et al. (1992) sluiced gravel specimens with sand. It was believed that sluicing would reduce membrane compliance without significantly changing the noncompliant cyclic loading resistance of the gravel. Comparison of the cyclic loading resistance of sluiced and unsluiced 9.5-mm by 4.75-mm gravel specimens at a relative density of 58% is shown in Fig. 2. It may be seen that the cyclic loading resistance of the sluiced gravel is considerably lower than that for the unsluiced gravel. In fact, the sluiced specimens only had about 70% of the cyclic loading resistance of the unsluiced specimens. Thus, to account for the effects of membrane compliance in such specimens, only 70% of the cyclic loading resistance determined by laboratory testing should be used as a basis for evaluating prototype performance.

Membrane Compliance Correction

The results of these tests and many other similar tests on other gravels were used to develop a correction for membrane compliance, shown in Fig. 3. This figure was developed for uniformly graded gravels, isotropically consolidated to about 200 kPa in 2.8-inch and 12-inch triaxial tests, and failing in approximately 10 to 30 stress cycles. The correction shown in the figure for 2.8-inch diameter specimens represents an average value developed from data presented by Evans and Seed (1987) and Martin et al. (1978). The noncompliant cyclic loading resistance may be determined by multiplying the compliant, laboratory determined cyclic loading resistance by the proposed correction factor. It should be noted that uniformly graded gravelly soils will experience significant membrane compliance effects while very well-graded gravelly soils will experience lesser membrane compliance effects. Therefore, the results of liquefaction tests performed on very well-graded gravelly soils tested in the triaxial test will require significantly smaller corrections for membrane compliance than those shown in Fig. 3.

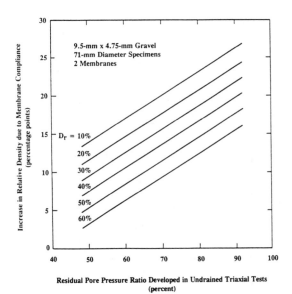

Figure 1.   Increase in relative density caused by membrane compliance (after Evans and Harder, 1993)

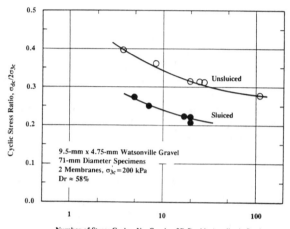

Number of Stress Cycles, N_c, Causing 5% Double Amplitude Strain

Figure 2. Cyclic loading resistance of sluiced and unsluiced gravel specimens (after Evans et al., 1992)

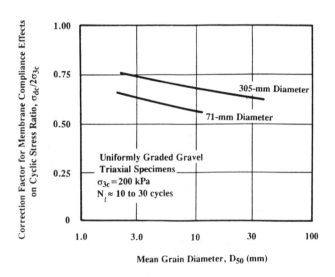

Mean Grain Diameter, D_{50} (mm)

Figure 3. Correction factor for membrane compliance effects (after Evans and Harder, 1993)

LABORATORY CYCLIC STRENGTH OF SAND-GRAVEL COMPOSITES

Introduction

Since most liquefied natural soil deposits are sand or sand-gravel composites, it was considered necessary to quantify the effect of gravel content on the liquefaction resistance of sand-gravel composites. This study presents preliminary results of a series of undrained cyclic triaxial tests performed on sand-gravel composite specimens with gravel contents of 0%, 20%, 40%, and 60%. The gravel and sand used in this study were obtained from commercial suppliers. Grain size distribution curves are shown in Fig. 4, and grain size parameters are listed in Table 1. All triaxial test specimens were about 71 mm in diameter, constructed in 5 equal mass layers by pluviating the sand-gravel composite through air. Material segregation was not noted for these sand -- gravel proportions. Test specimens were confined with 2 latex rubber membranes, each 0.3-mm thick.

Table 1. Grain Size Parameters

Material	D_{60} (mm)	D_{50} (mm)	D_{20} (mm)	D_{10} (mm)	C_u	G_s
Sand A	0.45	0.40	0.28	0.23	2.0	2.63
Sand B	0.27	0.24	0.17	0.11	2.6	2.57
Gravel A	7.00	6.50	5.60	5.00	1.4	2.72

Influence of Gravel Content

Fig. 5 shows a plot of cyclic stress ratio causing 5% double amplitude strain in 10 cycles, CSR $_{5\%-10}$, versus void ratio for Sand A specimens with 0%, 20, 40%, and 60% gravel content. The liquefaction resistance of sand-gravel composites may be seen to increase significantly with increasing gravel content. The value of CSR $_{5\%-10}$ approximately doubles with increasing gravel content from 0% to 60%. Values of CSR $_{5\%-10}$ are not corrected for membrane compliance because membrane penetration during hydrostatic loading was negligible. Thus, it was considered that membrane compliance effects during undrained cyclic loading would also be negligible. The increase in CSR$_{5\%-10}$ with gravel content is due to a significant drop in composite void ratio, as may be seen in Fig. 5. It may also be seen that the specimen void ratio increases with increasing gravel content even though the composite specimen relative density remains constant at 40%.

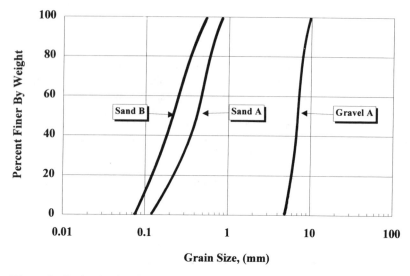

Figure 4. Grain size distribution curves for the materials used in this study

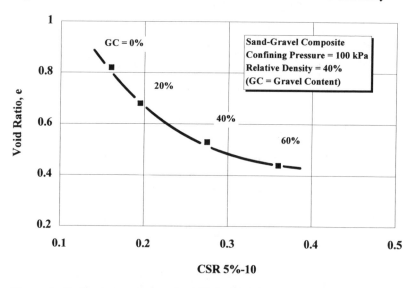

**Figure 5. Cyclic stress ratio causing 5% double amplitude strain in 10
cycles, CSR 5%-10, versus void ratio for sand A specimens
with 0%, 20%, 40%, and 60% gravel**

When the gravel content is less than about 40%, gravel particles can be considered to float in a sand matrix. The average density of the matrix sand in sand-gravel composite specimens may be determined by dividing the mass of the matrix sand by the matrix volume. For composite specimens at 40% relative density, the relative density of the matrix sand was determined to be 40%, 33.6%, 28.5% and -6%, respectively, for gravel contents of 0%, 20%, 40%, and 60%, as shown in Table 2. Thus the average matrix sand density decreases with increasing gravel content. The test results indicate that the cyclic loading resistance of sand-gravel composites increases even if the density of the sand matrix decreases. This implies that the cyclic loading resistance of sand-gravel composites may be controlled by overall composite specimen characteristics rather than matrix characteristics even though gravel particles are floating in the sand matrix.

Table 2 summarizes test data and material properties for sand-gravel composite specimens with gravel contents of 0% to 60%. Rows 4 and 5 show that, although the relative density of the composite is constant at 40%, the relative density of the matrix sand decreased significantly in this study with increasing gravel content. Similarly, row 7 shows that $CSR_{5\%-10}$ approximately doubles from 0.16 to 0.36 with increasing gravel content.

Table 2. Composite Material Properties

Row No.	Description	Sand A	Sand A --Gravel Composite			
(1)	Percent Gravel	0	20	40	60	100
(2)	Maximum Dry Density (Mg/m^3)	1.67	1.75	1.92	2.03	1.62
(3)	Minimum Dry Density (Mg/m^3)	1.33	1.48	1.66	1.79	1.39
(4)	Relative Density, composite, %	40	40	40	40	--
(5)	Average Relative Density, Matrix,	40	33.6	28.5	-6	--
(6)	Density, comp., dry (Mg/m^3)	1.45	1.57	1.75	1.87	--
(7)	CSR 5%-10, cyclic stress ratio, 5% strain in 10 cycles	0.16	0.195	0.275	0.36	--
(8)	CSR comp/CSR sand	1.00	1.22	1.72	2.25	--
(9)	Behavioral Relative Density, % (40%*CSR 5%-10)	40	49	69	90	--

Comparison with Dense Sand Behavior

Based on the increased $CSR_{5\%-10}$ shown in Fig. 5, one might expect the relative density of the matrix sand to increase as shown in Table 2, row 9. The values in row 9 were estimated by multiplying the normalized cyclic stress ratio (row 8) by 40% relative density. That is, sand specimens at the relative densities shown in row 9 would be expected to have approximately the same $CSR_{5\%-10}$ values shown in row 7. This comparison shows that a 40% relative density sand-gravel composite with 40% gravel may have about the same $CSR_{5\%-10}$ value as 69% relative density sand. (It should be noted that this comparison applies only to the materials tested.)

Plots of excess pore pressure and axial strain versus time are shown in Fig. 6(a) and 6(b), respectively, for 40% and 65% relative density sand, and 40% relative density sand-gravel composite with 40% gravel content. It may be seen that the 40% relative density sand reached a pore pressure ratio of almost unity in the 10th load cycle. During each successive load cycle, the pore pressure varied from a high value of about unity at zero axial strain to a low value of about 0.8 at peak axial strain. Before the 10th load cycle, pore pressure was building gradually but no appreciable strain developed. Double amplitude strain reached a value of 5% in the 10th cycle and exceeded 10% by the 12th cycle. This illustrates classic liquefaction behavior of loose to moderate density sand.

It may also be seen in Fig. 6 that the cyclic behavior of the 40% relative density sand-gravel composite with 40% gravel was similar to the 65% relative density sand. Pore pressure developed more rapidly in these specimens because greater cyclic stresses were applied. Pore pressure ratio reached 80% around the fourth cycle. Axial strain increased gradually and steadily from the application of the first few load cycles. Double amplitude strain reached 5% (our failure criterion) in about 10 load cycles, however, this value was not significantly exceeded even out to 15 cycles. During each load cycle, the pore pressure ratio varied from a high value of about 0.95 at zero axial strain to a low value of about 0.4 at peak axial strain. This significant drop in pore pressure, caused by specimen dilation during extension, results in specimens resistant to deformation. This is classic cyclic mobility behavior exhibited by dense sand. Thus, the loose sand-gravel composite (Dr=40%) with 40% gravel behaves like a dense sand.

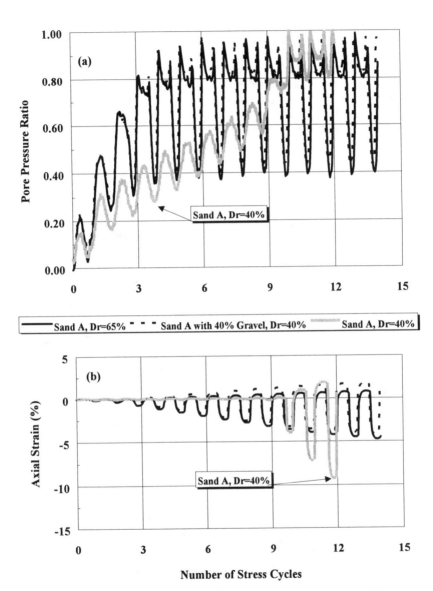

Figure 6. (a) Excess pore pressure and (b) axial strain versus number of stress cycles for 40% and 65% relative density sand, and 40% relative density sand-gravel composites with 40% gravel

Fig. 7 shows a comparison of the CSR causing 5% double amplitude strain or 80% pore pressure ratio versus number of stress cycles for 40% and 65% relative density sand and 40% relative density sand-gravel composite with 40% gravel. It may be seen that the cyclic loading resistance of the 40% relative density composite is nearly the same as the 65% relative density sand. The difference between the number of stress cycles causing 5% double amplitude strain and causing 80% pore pressure ratio is small for the loose sand but considerably greater for the dense sand and the composite specimen.

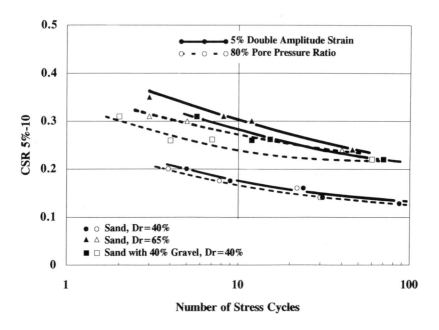

Figure 7. Cyclic stress ratio causing 5% double amplitude strain or 80% pore pressure ratio versus number of stress cycles for 40% and 65% relative density sand, and 40% relative density sand-gravel

Stress -- strain hysteresis loops are shown in Fig. 8a and 8b for 40% and 65% relative density sand, and 40% relative density sand-gravel composite with 40% gravel. Shown are the first and fifth stress cycles for specimens reaching 5% double amplitude strain in about 10 stress cycles. It may be seen that the 40% relative density composite specimen behaves more like the dense sand (Dr = 65%) than like the loose sand (Dr = 40%). The strain, stress, modulus, and damping values are nearly identical for the 40% relative density sand-gravel composite and the 65% relative density sand. The 40% relative density sand, on the other hand, has a lower cyclic resistance, and a considerably higher modulus and lower strain and damping at the fifth stress cycle.

The data presented above should clearly indicate the inaccuracies in scalping oversized particles from gap-graded soils and testing the matrix material at the same relative density as the in situ prototype. A more reasonable estimate may be obtained by testing the matrix material at a greater relative density to account for the increased cyclic loading resistance caused by the gravel inclusions. Additional study is being performed to validate this approach for a wide range of soils.

Correlation Between Liquefaction Resistance and Normalized Void Ratio

For any granular soil, it has been shown that reasonably good correlation may be obtained between cyclic loading resistance and relative density. However, when soils vary in gradation, e_{max}, e_{min}, and gravel content; void ratio or relative density may not correlate well with cyclic strength. A plot of $CSR_{5\%-10}$ versus relative density is shown in Fig. 9a for the sand-gravel composites tested in this study. It may be seen that relative density is not a good indicator of the cyclic loading resistance of these specimens. Thus, another soil parameter should be investigated to correlate with $CSR_{5\%-10}$ in gravelly soils.

A plot of $CSR_{5\%-10}$ versus void ratio is shown in Fig. 9b for the soils tested in this study along with data from other investigators (see Table 3). It may be seen that the composite specimens at 40% relative density show reasonably good correlation between $CSR_{5\%-10}$ and void ratio. This relationship may be improved by plotting $CSR_{5\%-10}$ versus normalized void ratio function, $F(e)/F(e_{min})$. The normalized void ratio function was used successfully by Tokimatsu and Uchida (1990) to determine normalized shear modulus, G_N. Good correlation was found between liquefaction resistance and normalized shear modulus for soils having different ranges of void ratio. The normalized shear modulus is expressed as follows:

$$G_N = AF(e)/F(e_{min})$$

where A is a parameter reflecting the effect of soil fabric, and $F(e)/F(e_{min})$ is a normalized void ratio function where:

$$F(e) = (2.17-e)^2/(1+e).$$

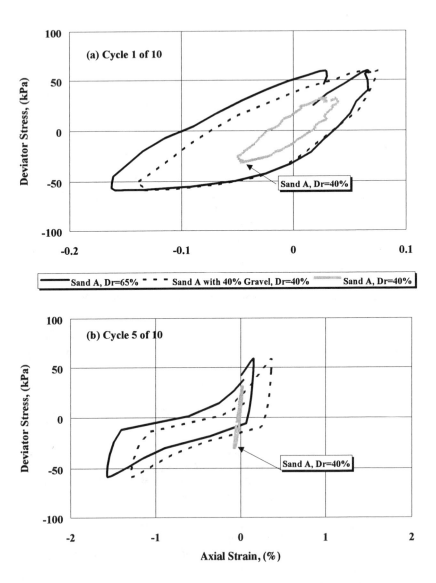

Figure 8. Stress -- strain hysteresis loops for 40% and 65% relative density sand, and 40% relative density sand-gravel composites with 40% gravel (a) first cycle of ten cycles, (b) fifth cycle of ten cycles

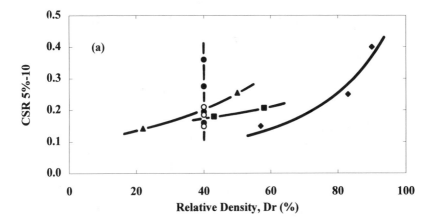

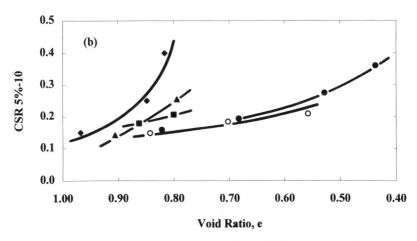

Figure 9. Cyclic stress ratio causing 5% double amplitude strain in 10 cycles, CSR 5%-10, versus (a) relative density and (b) void ratio for various sand-gravel composites

Table 3 — Composite Material Properties

Soil	D$_{50}$ (mm)	D$_r$ (%)	e$_{max}$	e$_{min}$	e	F(e)/F(emin)	CSR 5%-10	Reference
Sand A	0.4	40	0.982	0.580	0.821	0.624	0.160	This Study
with 20% Gravel		40	0.795	0.514	0.683	0.726	0.195	
with 40% Gravel		40	0.616	0.395	0.528	0.782	0.275	
with 60% Gravel		40	0.509	0.328	0.437	0.819	0.360	
Sand B	0.2	40	1.000	0.605	0.842	0.627	0.155	
with 20% Gravel		40	0.833	0.505	0.702	0.688	0.185	
with 40% Gravel		40	0.656	0.408	0.557	0.758	0.210	
3/4"x1-1/2"	27.0	22	0.992	0.597	0.905	0.542	0.143	Evans and Seed (1987)
Watsonville		50	0.992	0.597	0.795	0.680	0.255	
3/8"x#4	6.5	43	1.039	0.627	0.862	0.628	0.180	
Watsonville		58	1.039	0.627	0.800	0.713	0.207	
Niigata Sand	0.2	57	1.230	0.770	0.968	0.663	0.150	Tokimatsu (1990)
		83	1.230	0.770	0.848	0.854	0.250	
		90	1.230	0.770	0.816	0.912	0.400	

For reconstituted laboratory specimens, the parameter A is related to the method of specimen preparation. If the same method is used for specimen preparation in testing different soils, parameter A should be approximately constant. Therefore, $CSR_{5\%-10}$ may be well correlated with $F(e)/F(e_{min})$. Additional data was assembled to test the hypothesis that $CSR_{5\%-10}$ could be correlated with normalized void ratio function for different soils. Fig. 10 shows a plot of $CSR_{5\%-10}$ versus normalized void ratio function for the data in Table 3. It may be seen that there is good correlation between these two parameters for a specific soil. However, there is significant scatter about the mean when all the data is considered. Thus, in its present form, normalized void ratio function may not accurately predict cyclic loading resistance for a wide range of soils. However, with additional study ongoing in this area, this relationship may be modified to provide a reasonable estimate of the cyclic loading resistance of gravelly soils.

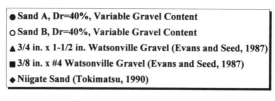

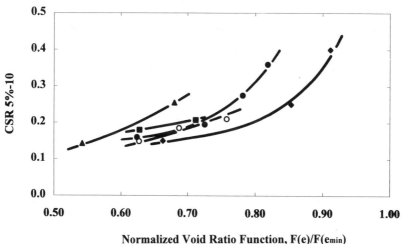

Figure 10. Cyclic stress ratio causing 5% double amplitude strain in 10 cycles, CSR 5%-10, versus normalized void ratio function, $F(e)/F(e_{min})$, for various sand-gravel composites

CONCLUSIONS

Membrane compliance may cause pore water redistribution during undrained testing of gravel, leading to significantly increased density at the end of the test. Volume changes measured during drained hydrostatic rebound may be used to compute changes in specimen density that are expected to occur during undrained cyclic loading. Relative density may increase by up to 20 percentage points or more during undrained cyclic loading at confining pressures in the range of 200 kPa due to membrane compliance.

Sluicing may significantly reduce the effects of membrane compliance in undrained, cyclic triaxial tests performed on coarse- or uniformly-graded gravel specimens. The results of this investigation indicate that a reasonably accurate assessment of the noncompliant cyclic loading resistance of the gravels tested in this study may be determined by testing sluiced specimens or applying the correction presented.

The liquefaction resistance of sand-gravel composites may increased considerably with increasing gravel content, even for composites of the same relative density. It was found that at 40% gravel content, a 40% relative density composite will behave like a 65% relative density sand. This considerable increase in liquefaction resistance of sand-gravel composites was not caused by membrane compliance, but may be attributed to the decrease in void ratio of sand-gravel composites with the inclusion of gravel particles. The effect of gravel content on the liquefaction resistance of sand-gravel composites can be evaluated by considering the increase in density of the composite materials.

Scalping oversized particles from gap-graded soils will underestimate the cyclic strength if the matrix is tested at the same relative density as the prototype. A more reasonable estimate may be obtained by testing the matrix material at a greater relative density to account for the increased cyclic loading resistance caused by the gravel inclusions. Additional study is being performed to validate this approach for a wide range of soils.

Based on laboratory-determined liquefaction resistance data from sand-gravel composites at various gravel contents, it was found liquefaction resistance is well correlated with normalized void ratio function for a given soil. The correlation indicated significant data scatter about the mean when data from a wide range of soils is considered. This scatter may be due to aging, stress history, specimen reconstitution method, and other factors. Additional research, especially field performance, is needed to validate and improve the proposed correlation.

ACKNOWLEDGMENTS

This investigation was funded, in part, by Grant No. MSS-9110153 from the National Science Foundation. The support of the Foundation and Mehmet Tumay is greatly appreciated.

APPENDIX I -- REFERENCES

Banerjee, N.G., Seed, H.B., and Chan, C.K. (1979). "Cyclic behavior of dense coarse-grained materials in relation to the seismic stability of dams." Report No. UCB/EERC-79/13, Earthquake Engrg. Res. Ctr., College of Engrg., Univ. of California, Berkeley, Calif.

Evans, M.D. and Harder, L.F. (1993). "Liquefaction potential of gravelly soils in dams." Geotechnical Practice in Dam Rehabilitation, ASCE, Geotechnical Special Publication No.35, 467-481.

Evans, Mark D., "Density Changes During Undrained Loading - Membrane Compliance", Journal of Geotechnical Engineering, ASCE, Vol. 118, No. 12, December, 1992.

Evans, Mark D., Seed, H.B. and Seed, R.B., "Membrane Compliance and Liquefaction of Sluiced Gravel Specimens", Journal of Geotechnical Engineering, ASCE, Vol. 118, No. 6, June 1992.

Evans, M. D. and Seed H. B. (1987), " Undrained Cyclic Triaxial Testing of Gravels - The Effect of Membrane Compliance", Report No. UCB/EERC-87/08,Earthquake Engineering Research Center, College of Engineering, Univ. of California, Berkeley, Calif.

Harder, L.F., Seed, H.B. (1986), "Determination of Penetration Resistance for Coarse-Grained Soils Using the Becker Hammer Drill," Earthquake Engineering Research Center, Report No. UCB/EERC-86/06, May, 1986.

Hardin,B.O., and Richart, F.E. (1963)."Elastic wave velocity in granular soils." J. Soil Mechanics and Foundation Engrg., ASCE, 89(1),33-65.

Hynes, M.E. (1988)."Pore Pressure Generation Characteristics Of Gravel Under Undrained Cyclic Loading." Ph.D thesis, Univ. of California, Berkeley, Calif.

Martin, G.R., Finn, W.O.L., and Seed, H.B. (1978) "Effects of System Compliance on Liquefaction Tests," Journal of Geotechnical Engineering, ASCE, 104(4).

Nicholson, P.G., Seed, R.B., and Anwar, H.A. (1993a) "Elimination of Membrane Compliance in Undrained Triaxial Testing. I. Measurement and Evaluation", Cannadian Geotechnical Journal, vol. 30, p 727 - 738.

Nicholson, P.G., Seed, R.B., and Anwar, H.A. (1993b) "Elimination of Membrane Compliance in Undrained Triaxial Testing. II. Mitigation by Injection Compensation", Cannadian Geotechnical Journal, vol. 30, p 739 - 746.

Seed,R.B., Anwar, H.A., and Nicholson, P.G. (1989)."Elimination of membrane compliance effects in undrained testing of gravelly soils." Proc. of the 12th Int. Conf. on Soil Mechanics and Foundation Engrg., Rotterdam, Netherlands, 111-114.

Tokimatsu, K., and Uchida, A. (1990)."Correlation between liquefaction resistance and shear wave velocity." Soil and Foundation, Vol.30,No.2,33-42.

Wong, R.T., seed, H.B., and Chan, C.K. (1974). "Liquefaction of gravelly soils under cyclic loading conditions." Report No. UCB/EERC-74/11, Earthquake Engineering Research Center, College of Engineering, Univ. of California, Berkeley, Calif.

APPENDIX II. -- NOTATION

A -- a parameter reflecting the effect of soil fabric

C_u -- coefficient of uniformity, D_{60}/D_{10}

D_r -- relative density

D_{60} -- particle diameter for 60% finer by weight

D_{50} -- particle diameter for 50% finer by weight

D_{20} -- particle diameter for 20% finer by weight

D_{10} -- particle diameter for 10% finer by weight

e -- void ratio

$e_{max, min}$ -- maximum and minimum void ratio

G_N -- normalized shear modulus

G_s -- specific gravity of solids

r_u -- residual pore pressure ratio

$\gamma_{dmax, min}$ - maximum and minimum dry density

σ_{3c} - effective minor principal stress at consolidation

σ_d - cyclic deviator stress

$CSR_{5\%-10}$ -- Cyclic stress ratio ($\sigma_d/2\sigma_{3c}$) causing 5% double amplitude strain in 10 stress cycles

$F(e)/F(e_{min})$ -- normalized void ratio function

Gravelly Soil Properties Evaluation
by Large Scale In-situ Cyclic Shear Tests

Takaaki Konno [1], Munenori Hatanaka [2]
Kenji Ishihara [3], Yukimi Ibe [4], Setsuo Iizuka [5]

ABSTRACT

In order to verify the seismic stability of gravelly soil layers, a series of field cyclic loading tests were performed using a concrete block with an equivalent earth contact pressure of actual reactor building. In addition, a series of laboratory tests using a large scale triaxial test apparatus on high quality undisturbed gravel samples obtained by in-situ freezing method and large scale in-situ soil column tests using two soil columns with 10m in diameter, were performed for evaluation of the gravelly soil properties under seismic condition.

1 INTRODUCTION

The basic policy in Japan is to build nuclear reactor buildings on rock. But, in order to cope with the middle and long term siting problems it has become necessary to develop new siting technology from the standpoint of expanding the available range of site selections and effective utilization of lands. The gravelly soil layer in the Quaternary deposits has high possibility of becoming the bearing soil stratum when constructing a nuclear power plant.

In order to verify the seismic stability of such gravelly soil layers, a series of field cyclic loading tests were performed using a concrete block weighing 30MN with an earth contact pressure of 470kPa which is equivalent of actual reactor building, and a reaction block weighing approximately 50MN. The gravelly soil layer at the test site has a shear wave velocity of 380m/sec with SPT-N values of 40 to 50.

1.Manager, Kajima Corporation, Akasaka, Minato-ku, Tokyo, Japan
2.Chief Research Engineer, Takenaka Corporation, Ginza, Chuo-ku, Tokyo, Japan
3.Professor, University of Tokyo, Hongo, Bunkyo-ku, Tokyo, Japan
4.Deputy General Manager, 5.Manager, Nuclear Power Engineering Corporation, Toranomon, Minato-ku, Tokyo, Japan

In addition, a series of laboratory tests using a large scale triaxial test apparatus on high quality undisturbed gravel samples obtained by in-situ freezing method which is considered the best in the present state-of-the art technique and large scale in-situ soil column tests using two soil columns with 10m in diameter but with different depths of 5m and 9m, were performed for evaluation of in-situ gravelly soil properties used in the simulation analyses of concrete block tests. Fig. 1 shows the general view of the field test models. In the field cyclic loading tests, the cyclic deformation characteristics of the gravel layer were investigated to large shear strain range exceeding 10^{-3}. The excess pore water pressure during cyclic shear was very small, which is well corresponding to the high undrained cyclic shear strength obtained in the laboratory test on undisturbed samples. And the non-linear behavior of the concrete block such as hysteresis loops of vertical displacement of the soil, tilt of the concrete block, and lateral displacement between the concrete block and the reaction block were observed. These test results were simulated by 2-dimensional FEM non-linear analyses using the soil properties evaluated by in-situ soil column tests and laboratory tests on in-situ freezing samples.

The objectives of this paper are to present the gravelly soil properties under seismic condition evaluated in cyclic loading tests of the concrete block, and from both laboratory tests on high-quality undisturbed samples recovered by in-situ freezing method and the in-situ large scale soil column tests.

Fig.1 General view of the field test model

2 CYCLIC LOADING TESTS OF CONCRETE BLOCK

2.1 Test site construction

The test site was selected in the field of Tadotsu Engineering Laboratory of NUPEC, in Kagawa Prefecture, Japan, because the test site has a diluvial gravel layer underlying 11m below the ground surface. The surface reclamation layer of dredged material was excavated to the depth of 11m below the ground surface to expose the underlying gravel layer and then concrete blocks and in-situ soil columns were constructed. The ground water level was lowered and controlled to hold at the distance of -1.5m from the excavated ground surface by pumping up during the field tests.

The concrete block used for this tests was designed to have the dimensions in plan 8m × 8m at the lower part and 12m × 12m at the upper part and 10m height so that the earth contact pressure of approximately 470kPa that is equivalent to actual reactor building could be attained. As a reaction block in cyclic loading tests, a rectangular concrete block with a dimension of 16.5m × 16.5m × 8m was constructed beside the concrete block. To observe and record the gravelly soil behavior during the tests, many measuring instruments were installed in the test ground before construction of the concrete block and the reaction block as shown in Fig.2.

The items measured and the measuring instruments were 1) Vertical displacement of the ground by settlement gauges, 2) Tilt of the concrete block and the soil by tilt gauges, 3) Relative lateral displacement between the concrete block and the reaction block by lateral displacement gauges, 4) Sliding displacement between the concrete block and the bearing soil layer by displacement gauges and 5) Excess pore water pressure in the ground by pore water pressure gauges.

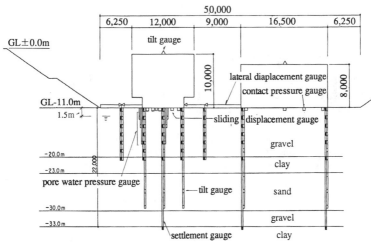

Fig.2 Section of the field test site and the array of measuring instruments

2.2 Soil profile at the test site

Shown in Fig. 3 are typical soil layer composition of the test site, penetration resistance value (N and NL) of the standard penetration test (SPT) and the large scale penetration test (LPT), and the distribution of the shear wave velocity through the depth obtained by the down-hole method. The test site is composed of reclaimed soil of dredged material 11m in thickness from the ground surface, underlaid by a

diluvial gravelly soil layer from the depth of 11m to 20m. This gravelly soil layer has
a shear wave velocity of 380m/sec with N values of 40 to 50. The value of NL in
this gravelly layer is between 15 and 40. According to the in-situ freezing sampling
method, 20 samples each 30cm in diameter and 60cm long, were extracted from the
gravelly soil layer in the test site to obtain the gravelly soil properties by laboratory
tests.

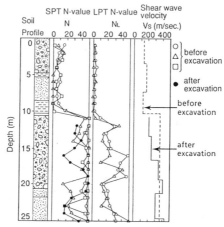

Fig.3 Soil profile of test site

2.3 Loading procedures for concrete block tests

  Horizontal cyclic load was applied by a pair of hydraulic jacks (maximum
loading limit is 6MN/jack) at the center of the gravity of the concrete block in both
push and pull directions. For the cyclic loading tests, the target shear strain level of
the soil was 10^{-3}. The loads were applied in both push and pull directions at the
speed of 1.18MN/min, and the loading level was increased in 10 steps of 980kN
increment up to the maximum 9.8MN. At each loading step the cyclic loading was
repeated 5 times, as shown in Fig.4.

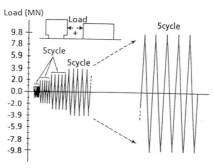

Fig.4 Loading Pattern by hydraulic jacks

2.4 Results of cyclic loading tests

The deformation characteristics of the gravelly soil showed non-linearity starting from the small strains, and residual deformation increased as the cyclic load increased. The test results of vertical displacement of the soil, tilt of the concrete block, and lateral displacement between the concrete block and the reaction block are shown in Fig.5 - Fig.9. In these figures, the hysteresis loops from 980kN to 9.8MN loading at only the fifth cycle of each loading step are shown for clarity.

Some remarks obtained from the test results are as follows:

1) The vertical displacement of the soil; The hysteresis loops of the vertical displacements showed different shape at the place of the edge and the center of the concrete block, as shown in Fig.5. They were caused by the combination of the normal stress and the shear stress at the edge, on the other hand by mainly shear stress at the center respectively. The residual displacement by cyclic loading increased as the load increase and the rate of increasing of the residual displacement showed the tendency of becoming large even in 9.8MN loading step, as shown in Fig.6.

The maximum residual displacement is 31mm at the center of concrete block in 9.8MN loading.

The shear strain was obtained from tilt gauge measurement and its magnitude exceeded 10^{-2} at near the corner of the concrete block, and exceeded 10^{-3} at wide area below the concrete block at the loading step of 9.8 MN, as shown in Fig.7.

2) The tilt of the concrete block; As the load increased in excess of 4.9MN, the increase of the rate of the tilt amplitude became noticeable. Then, a residual tilt occurred in the direction of pull load, as shown in Fig. 8.

3) The relative lateral displacement between the concrete block and the reaction block; The amplitude increased as load increased, and residual lateral displacement occurred in the direction of pull load, as shown in Fig. 9.

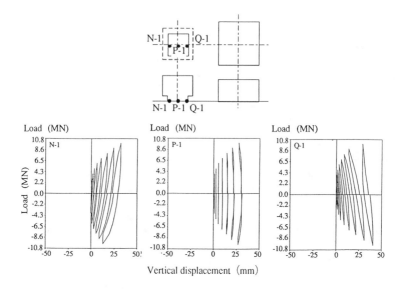

Fig.5 Hysteresis loops of vertical displacement of soil
Hysteresis loops shown are only fifth cycle of each loading level.

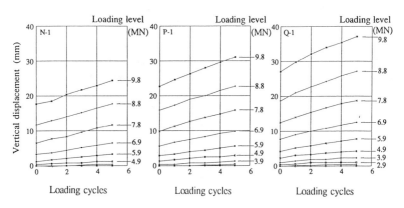

Fig.6 Relationships between loading cycles and vertical displacement of the soil
In these relationships, N-1 represents the displacement at fifth push loading,
P-1 represents the displacement at fifth unloading, Q-1 represents the
displacement at pull loading.

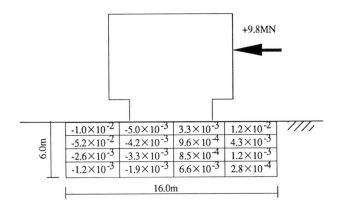

Fig.7   Shear strain of the soil under concrete block

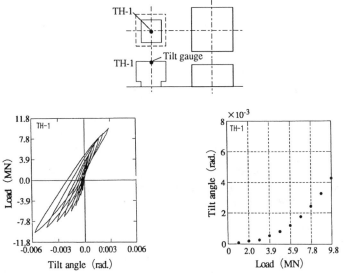

(a) Hysteresis loops of tilt angle. Only the
5th cycle loops of each loading level are
shown.

(b)Relationships of load and tilt angle.
The tilt angle represent a half of double
amplitude at each 5th cycle load.

Fig.8   Tilt angle of concrete block

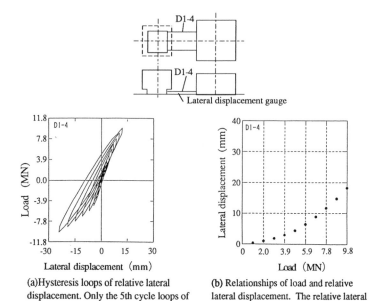

(a)Hysteresis loops of relative lateral displacement. Only the 5th cycle loops of each loading level are shown.

(b) Relationships of load and relative lateral displacement. The relative lateral displacement represents a half of double amplitude at each 5th cycle load.

Fig.9   Relative lateral displacement between concrete block and reaction block

3 EVALUATION OF GRAVELLY SOIL PROPERTIES

The engineering properties of the gravelly soil layers in the test site were evaluated both by laboratory tests and in-situ large scale soil column tests. The high-quality undisturbed gravel samples were recovered by in-situ freezing sampling method, the details of the sampling procedure is reported by Suzuki et al. (1992). Fig.10 shows an example of gravel sample with 30cm in diameter and 60cm in height. Perfect smoothness of both ends and cylindrical surface of the sample can be seen. The samples have an average grain size of $3-17$mm, with a maximum grain size of $70-150$mm. Fine content is 0 - 3%, and dry density is $17.8-20.0$ kN/m^3, and the specific gravity is $2.64-2.65$.

Four different types of laboratory tests were performed on undisturbed gravel samples. The specific laboratory test methods were; (1) undrained cyclic strength test (liquefaction test), (2) cyclic deformation test, (3) consolidated drained triaxial compression test, (4) isotropic compression test and isotropic expansion test. All of the laboratory tests were conducted using a large scale triaxial test apparatus as shown in Fig.11.

All the cyclic tests were conducted by applying uniform sinusoidal cycles of deviator stresses at a frequency of 0.01Hz. The low frequency was selected in order to maintain the constant deviator stress amplitude.

For the purpose of comparison, the undisturbed specimen was reconstituted following the completion of the experiment. Fundamentally, the granular composition and relative density of the reconstituted specimens were prepared to be the same as that of the undisturbed specimens.

 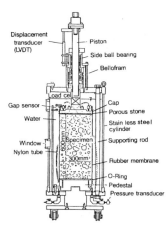

Fig.10 Undisturbed frozen gravel sample Fig.11 Large scale triaxial test apparatus

Test results are summarized in the following.

(1) Undrained cyclic strength (Cyclic mobility)

The undrained cyclic strength obtained by undrained cyclic triaxial test is shown in Fig.12. For comparison, the test results on undisturbed and reconstituted samples are shown. From the figure it can be seen that the undisturbed samples show a high value of cyclic shear stress ratio of 0.44 required to cause a double-amplitude axial strain of 2.5% in 20 cycles of load application. The strength of reconstituted samples is only about one half that of the undisturbed samples, and that the strength of the in-situ gravel soil cannot be correctly evaluated using the reconstituted soil.

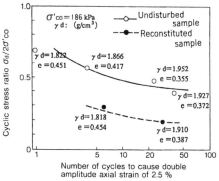

Fig.12 Undrained cyclic shear strength

Fig.13 shows the typical time histories of the cyclic deviator load, excess pore water pressure and axial displacement during cyclic strength tests for undisturbed samples. We recognized some characteristic points of difference in the excess pore water pressure and axial displacement. They are described below.

As shown in Fig.13(c), when the specimen stressed in extension side, the excess pore water pressure dropped to minus due to the large amount of dilatancy. Furthermore, the excess pore water pressure ratio did not reach 100%. These characteristics are like that of the undisturbed specimens of the dense sand reported by Yoshimi et al. (1984).

The increase in axial displacement due to the cyclic shear was gradual and appeared on the extension side from the early stage of shear for the undisturbed specimen (Fig. 13(b)). Fig.14 shows typical stress-strain relationships of the undisturbed specimen under the cyclic stress application. The stress strain relationships show the so-called reverse-S curves which resemble the stress-strain relationship observed for undisturbed dense sand as reported by Yoshimi et al. (1984).

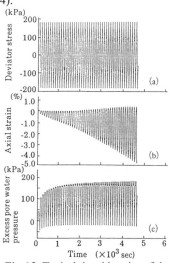

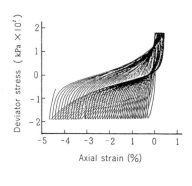

Fig.13 Typical time histories of the cyclic deviator load, axial strain and excess pore water pressure

Fig.14 Stress-strain relationship under cyclic undrained shear

(2) Cyclic deformation characteristics

①Strain dependency of shear modulus,G, and damping ratio, h.

The cyclic deformation characteristics obtained by cyclic undrained triaxial test is shown in Fig.15. For comparison, the results of reconstituted samples are also shown. The shear modulus, G, obtained by reconstituted samples is only about one half of that of the undisturbed samples, and as in the case with strength evaluation, it indicates that the deformation characteristics of the in-situ gravel soil cannot be evaluated by reconstituted samples.

Fig.15(c),(d) shows the G/Go~ γ relations for both undisturbed and reconstituted samples. In spite of the large difference of G~ γ relations between undisturbed sample and reconstituted sample, there is no significant difference in G/Go~ γ relation between them. Fig.16 shows similar data of G~ γ , h~ γ relations and also the G/Go~ γ relations on Tokyo gravel reported by Hatanaka et al . (1988). The same trend can be seen. From the simulation analysis performed in this study, the G~ γ , h~ γ relations were successfully modeled by the Hardin-Drnevich model.

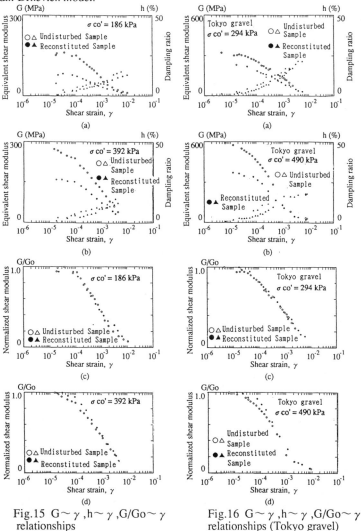

Fig.15 G~ γ ,h~ γ ,G/Go~ γ
relationships

Fig.16 G~ γ ,h~ γ ,G/Go~ γ
relationships (Tokyo gravel)

Fig.17 shows the $G/Go \sim \gamma / \gamma^{0.5}$ relations and $h \sim \gamma / \gamma^{0.5}$ relations, in which $\gamma^{0.5} = 0.5\%$ strain, obtained both from the laboratory tests and the in-situ soil column tests. There can be seen a good agreement of the $G/Go \sim \gamma / \gamma^{0.5}$ relations obtained from the laboratory tests and the in-situ soil column tests. The $h \sim \gamma / \gamma^{0.5}$ relations obtained from the in-situ soil column tests are scattered to some extent compared with those obtained from the laboratory tests. But the similar tendency that the damping ratio increases from about 2% to 15% with increasing shear strain can be seen.

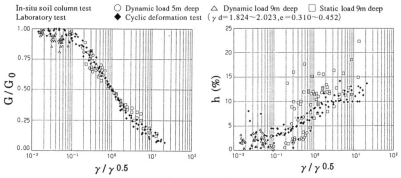

Fig.17 $G/Go \sim \gamma / \gamma^{0.5}$, $h \sim \gamma / \gamma^{0.5}$ relations obtained from
in-situ column test and laboratory test

② Shear modulus at minute strain
Fig.18 shows the relation of the shear modulus, Go, at small strain ($=10^{-5}$) and the confining stress σ_{co}'. On both logarithmic coordinates, an almost straight line relation is observed. The slope of the straight line is about 0.8, which is larger than the value of 0.5 commonly known for sand. This means that in case of the gravelly soils, the effect of the confining stress upon the shear modulus is more significant than that for the sand.

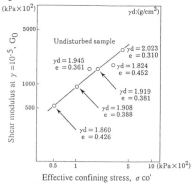

Fig.18 Relationships between Go and
σ_{co}' of undisturbed samples

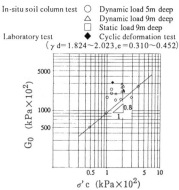

Fig.19 Comparison of initial shear
modulus obtained by three different
method

The shear modulus at minute strain, Go, obtained from three different methods (laboratory test, back analysis on the in-situ elastic wave test) are plotted in Fig.19 with the confining stress, σ_c', on both logarithmic coordinates. The following conclusions can be pointed out:

Among the three methods for the determination of Go for the same confining stress, the in-situ elastic wave method gives the largest value of Go, the laboratory test results show the smallest value, and the in-situ soil column tests provide a value between these two. The lower value of Go from laboratory test compared with that obtained from in-situ elastic wave test may be due to the effect of the bedding error as pointed out by Dong,J. et al. (1994).

In the initial stress analysis, the Go obtained in laboratory test on undisturbed samples was successfully used. However, in the simulation analysis on cyclic loading test of concrete block, Go from large scale in-situ column test gave better results.

(3) Static strength characteristics

The Mohr's circles at failure obtained by consolidated drained triaxial compression tests are shown in Fig.20. The internal friction angle of the undisturbed specimen is 36 to 37 degrees, and the cohesion is 25 to 67 kPa. The internal friction angle estimated from the N value (on the average, 43) using the empirical formula $\phi_d = \sqrt{12N} + 25$ proposed by Dunham for design purposes, is 48 degrees.

There is a big difference between the estimated value and the measured value on the undisturbed samples. In addition, there is some cohesion component that may be significant in some cases of actual design works.

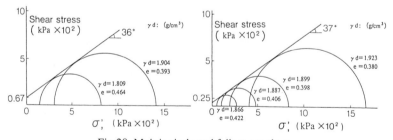

Fig.20 Mohr's circle and failure envelope

(4) Isotropic compression and expansion characteristics

For the same purposes as described in cyclic drained shear test, the coefficient of volume compressibility and the coefficient of volume expansion were measured using large scale triaxial test apparatus on undisturbed samples as follows.

After the undisturbed sample had been thawed and saturated, it was stressed isotropically under an initial effective stress of 19.6 kPa. After that the isotropic stress was increased to a certain level, it was reduced to the initial isotropic stress of 19.6 kPa. Then again it was increased to a certain value larger than that used in the former loading step.

Fig.21 shows an example of the relationship between the isotropic stress and the induced volumetric strain. From the Fig. 21, the coefficient of volume compressibility is ranging from 1.4×10^{-3} to 6.7×10^{-3}, and the coefficient of volume expansion is between 2.1×10^{-3} and 5.7×10^{-3}.

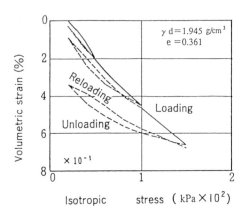

Fig.21 e-logp relationship under isotropic compression and expansion

4 SIMULATION ANALYSES OF CONCRETE BLOCK TESTS

4.1 Analytical method

Simulation analyses were performed using Hara-Kiyota model (Kiyota et al, 1988), and Nishi Model (Nishi et al, 1990). In this paper, the Hara-Kiyota Model was used in Analysis 1 and the Nishi Model was used in Analysis 2.

(1) Analysis 1

The Hara-Kiyota Model used in Analysis 1 is a method using non-linear shear stress-strain relationships and experimental equations for volumetric strain which explain well the initial shear modulus obtained from laboratory tests, decline of the soil stiffness and increase of the damping ratio. The amount of settlement is evaluated independently by using the time history of the shear strain of each soil element obtained in the response analysis to horizontal loading.

This model can express general stress condition in the soil as a combination of isotropic compression stress condition and a simple shear stress condition. A simple shear stress condition is evaluated by a type of Hardin-Drenevich model of the equation (1).

$$\frac{\tau}{\tau_r}=\frac{\gamma/\gamma_r}{1+|\gamma/\gamma_r|} \tag{1}$$

where, τ : shear stress, γ : shear strain, G_0 : initial shear modulus,
   τ_r : reference shear strength,
   γ_r : reference shear strain, $\gamma_r=\tau_r/G_0$

For the calculation of settlement, first, the volumetric strain is calculated by dividing into volumetric compression strain and volumetric expansion strain for each soil element and then, multiplying the thickness of the element to the volumetric strain using the equations (2) to (7).

$$\varepsilon_{vd}=\varepsilon_{vd}^-+\varepsilon_{vd}^+ \tag{2}$$

$$\varepsilon_{vd}^-=\widetilde{\varepsilon_{vd}^-}\cdot\left\{1-\exp\left(-\lambda\cdot\gamma_s^{\beta}\right)\right\} \tag{3}$$

$$\widetilde{\varepsilon_{vd}^-}=a_1\cdot\gamma_0^{b1} \tag{4}$$

$$\lambda=a_2\cdot\gamma_0^{b2} \tag{5}$$

$$\varepsilon_{vd}^+=\widetilde{\varepsilon_{vd}^+}\cdot a_3\cdot\left|\varepsilon_{vd}^-\Big/\widetilde{\varepsilon_{vd}^-}\right|^{b3}\cdot\left|\gamma\Big/\gamma_d\right|^{b4} \tag{6}$$

$$\widetilde{\varepsilon_{vd}^+}=a_4\cdot\gamma_0^{b5} \tag{7}$$

where, ε_{vd} : volumetric strain
   ε_{vd}^- : volumetric compression strain
   ε_{vd}^+ : volumetric expansion strain,
   $a_1\sim a_4, b_1\sim b_5, \beta$: material constants evaluated by laboratory tests

(2) Analysis 2

Nishi Model used in Analysis 2 is a method using constitutive relations incorporated with dilatancy effects based on elasto-plastic theory .

The hysteresis curves are defined as applying Masing's rule to the following hardening function equations (8) and (9).

$$\text{when } \eta^*\geq\eta_{max}^* \text{ ; } \sqrt{2I_2}=\frac{d\eta^*}{G_0^*\left(1-n^*/M_f\right)^2} \tag{8}$$

$$\text{when } \eta^*<\eta_{max}^* \text{ ; } \sqrt{2I_2}=\frac{d\eta_r^*}{G_0^*\left(1-\eta_r^*/2M_f\right)^2} \tag{9}$$

where, $\sqrt{2I_2}$: second invariant of plastic deviatoric strain tensor
   G_0^* : initial tangential shear modulus in the relation η^*
       and $\sqrt{2I_2}$
   M_f : relative stress ratio at failure

$\eta *, \eta *_r$: relative stress ratio, suffix r represents the stress
condition of the reversal point

$\eta *_{max}$: maximum value of $\eta *$ recieved in the past

The volumetric strain Vdf is evaluated by equation (11).

$$V_{df} = M^* \cdot \sigma_m \left(\eta_r^* \right)^n \tag{11}$$

where, V_{df} : final volumetric strain in cyclic shear deformation

m*,n : material constants (n=3 in triaxial compression test or
simple shear test)

4.2 Results of simulation analyses

(1) Analysis of initial stress condition
The soil stress condition before cyclic loading tests was calculated by means of the Duncan-Chang model which was used for analyses of the influence of excavation of the surface soil and construction of the concrete blocks. The FEM model is shown in Fig.22. The bottom of the model has a fixed condition, and horizontal displacement at both sides is restrained.

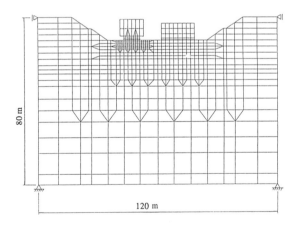

Fig.22 FEM analysis model

In initial stress condition analysis, the relations between the initial shear modulus and effective confining stress obtained from the results of laboratory test on in-situ soil frozen samples of the gravel layer was used as follows;

$$G_0 = 2350 \times \left(\sigma_{m0}' \right)^{0.8}$$

where, G_0 : initial shear modulus (kPa)

σ_{m0}' : effective confining stress (kPa)

The results of the initial stress condition analysis gave good agreement of the displacement of the concrete block between calculated value and measured value as shown in Fig.23 (a).

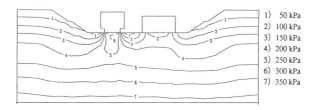

(a) Settlement after construction of concrete block and reaction block

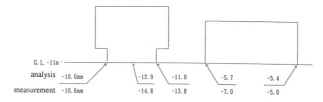

1) 50 kPa
2) 100 kPa
3) 150 kPa
4) 200 kPa
5) 250 kPa
6) 300 kPa
7) 350 kPa

(b) Confining stress (σ m) distribution

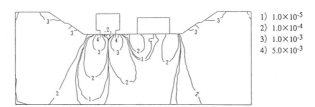

1) 1.0×10^{-5}
2) 1.0×10^{-4}
3) 1.0×10^{-3}
4) 5.0×10^{-3}

(c) Maximum shear strain distribution

Fig.23   Results of initial stress condition analysis

(2) Analysis of cyclic loading test

Simulation analyses with Analysis 1 and Analysis 2 were performed on the same loading pattern as the cyclic loading tests. The soil properties for the analyses were evaluated by the results of laboratory tests on undisturbed gravel samples and torsional tests on large in-situ soil columns as reported by Konno et al. (1993). In both analyses, the relations between the initial shear modulus and effective confining stress obtained from the laboratory cyclic undrained tests on in-situ frozen samples (Case 1), and from the large in-situ torsional soil columns test results of the gravel layer (Case 2) were used as follows:

$$\text{Case 1}: G_0 = 2350 \times (\sigma_{mo}')^{0.8}$$
$$\text{Case 2}: G_0 = 3800 \times (\sigma_{mo}')^{0.8}$$

where, G_0 : initial shear modulus (kPa)

σ_{mo}' : effective confining stress (kPa)

The FEM model is the same model used in initial stress condition analysis shown in Fig.22. The bottom of the model has a fixed condition, and vertical displacement at both sides is restrained. In Analysis 1, calculations were repeated until the stress-strain relation converges enough within each solution step. In Analysis 2, calculations were based on a load incremental method with a pitch of 19.6kN to ensure the accuracy of computation.

The results of simulation analyses, regarding the vertical displacement of the soil, the tilt of the concrete block, and the lateral relative displacement between the concrete block and the reaction block were shown compared with the measured one in Fig.24 - Fig.29.

The Hysteresis loops at 9.8MN cycle shows that Analysis 1 has a tendency of wider hysteresis loop than the measured one, because the material constant used in Analysis1 was provided by a laboratory test of Toyoura sand. Relationships of load and deformation show same tendency in the vertical displacement of soil between the analytical value and the measured one. The tilt of the concrete block and the lateral displacement between the concrete block and the reaction block show the same tendency with the measured one in Analysis 1, but Analysis 2 shows that the displacement increases with increasing load differently than the measured one. Both methods of analysis showed good agreement with the measured values when using the initial shear modulus evaluated by large in-situ soil column tests as shown in the analysis of Case 2.

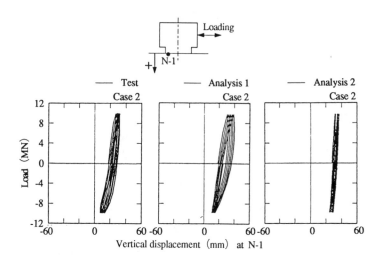

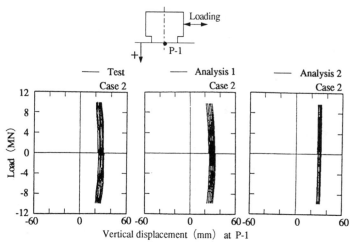

Fig.24 Hysteresis loops of the vertical displacement at loading step of 9.8MN

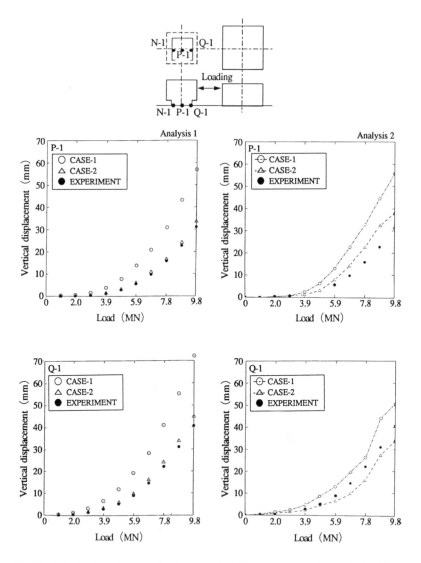

Fig.25 Relationships between load and vertical displacement.In these relationships,
        N-1 represents the displacement at fifth push loading,
        P-1 represents the displacement at fifth unloading,
        Q-1 represents the displacement at pull loading.

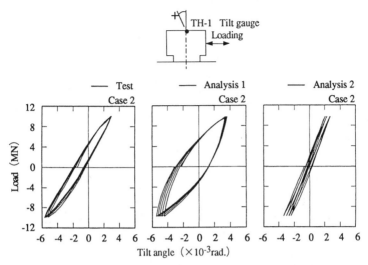

Fig.26 Hysteresis loops of the tilt angle of concrete block at loading step of 9.8MN

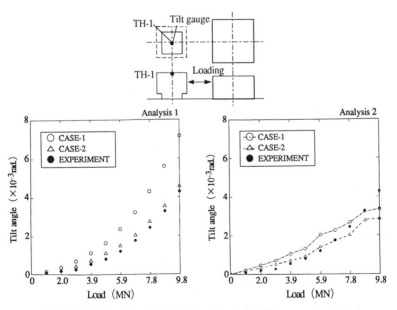

Fig.27 Relationship of the load and tilt angle of the concrete block
The tilt angle represent a half of double amplitude at each 5th cycle load.

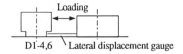

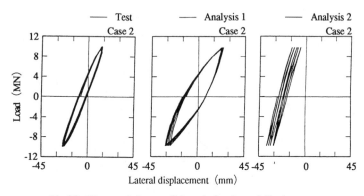

Fig.28 Hysteresis loops of the relative lateral displacement
between concrete block and reaction block

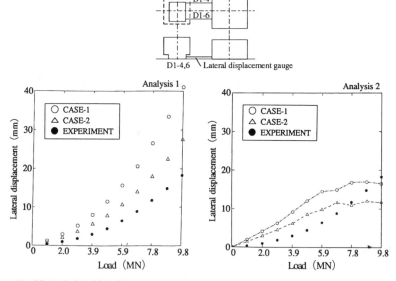

Fig.29 Relationship of load and relative lateral displacement between concrete
block and reaction block. The relative lateral displacement represents the half of
double amplitude at each 5th cycle load.

5 CONCLUSIONS

The following conclusions were drawn through the field cyclic shear tests and laboratory tests on undisturbed gravelly samples from a depth of 11m below the ground surface.

(1) The gravelly soil layer in the test site showed a large shear strain of 10^{-3} in the cyclic loading tests of the concrete block, but even in this shear strain level, the observed excess pore water was negligible and will not influence soil stability. The small pore water pressure is considered to be consistent with the high value of the undrained cyclic shear strength obtained in the laboratory test on undisturbed samples.

(2) The gravelly soil showed the volumetric deformation due to the dilatancy accumulated in the direction of compression with increasing number of loading cycle in the shear strain range larger than 10^{-3} in the concrete block test but it was not yet so large as compared with the final value of the displacement which was observed in laboratory test.

(3) The simulation analyses of the concrete block test gave the following results:
  a. The shear modulus at minute strain obtained from laboratory test on undisturbed gravel samples was used in initial soil stress analysis. Good agreement of the displacement after construction of the concrete block was obtained between calculated value and measured value.
  b. For the simulation analysis of the cyclic loading test of the concrete block, the shear modulus at minute strain obtained from the in-situ soil column tests gave good agreement of the displacement of the concrete block between calculated value and measured value.

(4) The simulation analyses of the concrete block tests showed that the correct evaluation of the soil properties is very important to estimate the non-linear response of a soil structure interaction in a large earthquake.

(5) Shear moduli at minute strains, Go, were obtained from three different methods. Among the three methods, the in-situ elastic wave method gave the largest value, the laboratory test showed the smallest value, and the in-situ soil column test provided a value between these two.

(6) Regarding the strain dependency of shear modulus and damping, there was no significant difference of the $G/Go \sim \gamma / \gamma^{0.5}$ relations and the $h \sim \gamma / \gamma^{0.5}$ relations between those obtained from the laboratory tests on undisturbed samples and those from the in-situ soil column tests.

(7) There was a significant difference of undrained cyclic shear strength between undisturbed and reconstituted gravel specimens. For example, the stress ratio required to cause double amplitude shear strain of 2.5% in 20 cycles of stress application for the undisturbed gravel specimens is about twice the value of the reconstituted specimens of the same gravel.

ACKNOWLEDGEMENTS

This work was carried out by NUPEC as the project sponsored by the Ministry of International Trade and Industry of Japan. This work was reviewed by Committee of Verification Test on Siting Technology for High Seismic Structures of NUPEC. The authors wish to express their gratitude for the cooperation and valuable suggestions given by every committee member.

REFERENCES

Dong, J., Nakamura, K., Tatsuoka, F. and Kohata, Y. 1994. "Triaxial behavior of gravels subjected to monotonic and cyclic loadings", Symposium on the testing method, field investigation method and the applicability of the test results - Deformation characteristics of geotechnical materials in the dynamic problems dealing with the soils and earth structures, Tokyo, pp.211~216.

Hatanaka,M., Suzuki, Y., Kawasaki, T. and Endo, M. 1988. "Cyclic undrained shear properties of high quality undisturbed Tokyo gravel", Soils and Foundations, Vol.28, No.4, pp.57~68.

Kiyota, Y., Hara, A., Sakai, Y. and Aoyagi, T. 1988. "Nonlinear dynamic response analysis by the 'deformation model of soils under combined stresses' ", 9WCEE,Technical Session : D4, Tokyo, Vol. III, pp.127~132.

Konno,T., Suzuki, Y., Tateishi, A., Ishihara, K., Akino, K. and Iizuka, S. 1993. "Gravelly soil properties by field and laboratory tests", Third International Conference on Case Histories in Geotechnical Engineering, St. Louis, Vol.I, pp.575~594.

Nishi,K. and Kanatani,M. 1990. "Constitutive relations for sand under cyclic loading based on elasto-plasticity theory", Soils and Foundations,Vol.30, No.2, 43~59.

Suzuki,Y., Hatanaka, M., Konno, T., Ishihara, K. and Akino, K. 1992. "Engineering properties of undisturbed gravel sample",Proc.10th World Conference on Earthquake Engineering,Madrid, Vol.3, pp.1281~1286.

Yoshimi,Y., Tokimatsu, K., Kaneko, O. and Makihara, Y. 1984. "Undrained cyclic strength of a dense Niigata sand",Soils and Foundations, Vol.24, No.4, pp.131~145.

BECKER TEST RESULTS FROM GRAVEL LIQUEFACTION SITES

By

Leslie F. Harder, Jr., M. ASCE[1]

ABSTRACT: The 1983 Borah Peak Earthquake (M_s=7.3) induced strong shaking in the Thousand Springs Valley of Central Idaho and was responsible for causing liquefaction of gravelly soils at several sites. Two of these sites, Pence Ranch and Whiskey Springs, provided excellent opportunities for applying the Becker Penetration Test for characterizing the liquefaction potential of the gravelly soils. This paper describes the application of the Becker Penetration Test at these two sites. The results of the investigation showed that these sites would have been predicted to liquefy using the Harder and Seed (1986) correlation between corrected Becker and SPT blowcounts, together with the Seed et al. (1985) correlation between SPT blowcount and liquefaction resistance. This gives good support for the use of the Becker Penetration Test in predicting future behavior at other gravel sites.

INTRODUCTION

Early on the morning of October 28, 1983, a large magnitude earthquake (M_s=7.3) occurred in Central Idaho. The epicenter of the earthquake was located near the base of Mount Borah in the Thousand Springs Valley. The strong shaking produced by the earthquake resulted in the development of liquefaction-induced lateral spreading at several locations. At two sites, the shaking was found to have caused lateral spreading from the liquefaction of gravelly soils. The two sites are:

1. The Pence Ranch.

2. The Whiskey Springs Slide.

---

[1]Supervising Engineer, California Dept. of Water Resources, P.O. Box 942836, Sacramento, CA 94236-0001.

The liquefaction distress at these sites was originally documented by Youd et al. (1985). Although sand and silt boils initially suggested that the soils which liquefied were sands and silts, it soon became clear that the liquefied soils were quite gravelly. Because documented cases of gravel liquefaction are relatively rare, these two sites provided an unusual opportunity to apply the Becker Penetration Test as a tool for investigating the liquefaction potential of gravelly soils. Accordingly, geotechnical investigations were performed at both sites using the Becker Penetration Test. At the Pence Ranch Site, Standard Penetration Test (SPT) tests were also performed during this study. Additional explorations at both sites using the SPT and the Cone Penetration Test (CPT) have also been carried out by Andrus et al. (1987).

THE 1983 BORAH PEAK EARTHQUAKE

The Borah Peak Earthquake occurred at approximately 8:06 a.m. local time on October 28, 1983. The earthquake was located in a rural area between the towns of Challis and Mackay, Idaho. The National Earthquake Information Service (NEIS) reported a surface wave magnitude (M_s) of 7.3 for the main shock. The earthquake resulted in the deaths of two children from falling building rubble in the town of Challis, and an estimated damage to buildings and property of approximately 12.5 million dollars.

The earthquake occurred on the Lost River Fault and propagated northwesterly from the epicenter in the Thousand Springs Valley. Fault plane solutions indicated a fault rupture zone dipping down approximately 50 degrees to the southwest. Investigations by Crone et al. (1985) showed that the surface expression of the fault rupture formed a Y-shaped pattern and extended up to 36 km in length (see Figure 1). Surface expressions of fault displacement were primarily normal slip accompanied by a minor left lateral component. The vertical offset of the fault at the surface was found to range generally between 1 and 2 m with a maximum measured offset of 2.7 m.

There were no seismographs available in the vicinity of the fault rupture at the time of the main shock. The nearest strong motion instruments were located between 88 and 109 km from the epicenter at the Idaho National Engineering Laboratory (INEL, see Figure 1). For stiff soil and rock sites where instruments were located in either basements or in the free field (nine instruments), the peak horizontal accelerations ranged from 0.02 to 0.08g.

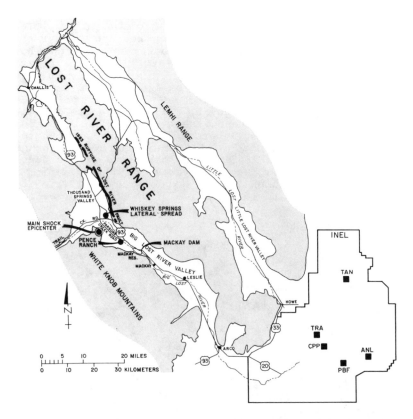

Figure 1:   Regional   Map   Showing   Geographic   Features
            Associated with the 1983 Borah Peak Earthquake
            (modified from Youd et al., 1985)

LIQUEFACTION OF GRAVELLY SOILS AT PENCE RANCH

General Description

        The Pence Ranch is located on a low terrace along
the Big Lost River approximately 8 km south of the
southern edge of the fault rupture (see Figure 1).
Harder (1988) employed damage intensity values and the
directivity of the propagation of the fault rupture to
estimate a peak ground acceleration of 0.29g at the Pence
Ranch Site.   According to the reconnaissance study
performed by Youd et al. (1985), a curved zone of
fissures and sand boils marked the head of a
liquefaction-induced lateral spread at this site.   Field
investigators found fissures that were as much as

30 cm wide and that had scarps as high as 30 cm.   The
head of the lateral spread passed beneath the ranch
house, through a steel barn, and out past adjacent hay
storage yards.

     Figure 2 shows a photograph of ejected sand that was
left after a sand boil erupted near the damaged ranch
house.   According to Youd et al. (1985), a sample of this
ejected sediment was found to be generally clean with a
gravel content of up to 5 percent (i.e. percent coarser
than the No. 4 sieve size).   Liquefaction related damage
at this site included differential movements of the house
foundation, sidewalks, and driveway.   In addition, a
small   concrete   pump   house   experienced   a   limited
floatation as evidenced by a heave of approximately 8 cm.

Figure 2:   Photograph of a Sand Deposit Left by a Sand
            Boil that Erupted Near the Pence Ranch House
            (Photograph taken November 1983 by T. L. Youd)

     Southeast of the ranch house, the fissures continued
through the bare dirt floor of a steel frame barn and out
through its doorway.   Continuing southeast, the fissures
ran through a hay storage yard and past a wire fence
enclosing the yard.   In this vicinity, the lateral
spreading caused the wire fence to pull approximately
75 cm apart.   This separation required the ranchers to
install a new fence post in the gap thus created, as
shown in Figure 3, to prevent cattle from entering the
yard.

Figure 3: Photograph of a Hay Yard Fence Pulled Apart by Lateral Spreading, Together with a Gravelly Sand Deposit Left after Eruption of Sand Boils (Photograph taken November 1983 by T. L. Youd)

The fissure that was formed where the wire fence separated, also shown in Figure 3, is of particular interest. At this location, a gravelly sand deposit was formed when sand, gravel, and water were ejected through the fissure. According to Youd et al. (1985), a sample of the ejected sediment was found to be generally clean with a gravel content of approximately 25 percent (i.e. percent coarser than the No. 4 sieve size). The ejected particles had sizes ranging up to 25 mm in diameter. Other sand boils in the vicinity were found to have significant gravel content as well.

1984 Drilling Explorations at Pence Ranch

Drilling explorations were performed in August 1984 at the Pence Ranch using a Becker AP-1000 drill rig. Both Becker Penetration and SPT testing were performed. A total of 4 plugged-bit Becker soundings, 4 open-bit Becker soundings, and 2 rotary SPT boreholes were performed at two sites on the Pence Ranch. The first site, Site A, was located near the broken wire fence at the hay storage yard. The other site, Site B, was located near the sand boil in front of the ranch house. The SPT tests were performed only at Site B. The locations of the soundings are shown in Figures 4 and 5.

The SPT penetration resistances measured at the Pence Ranch were not considered accurate as the significant gravel content commonly resulted in the split spoon sampler bouncing on and/or breaking gravel particles. In some cases, the SPT drive shoe was plugged with a large gravel particle, resulting in its being driven as a solid penetrometer. Details of the SPT explorations are described in Harder (1988) and are not discussed further herein.

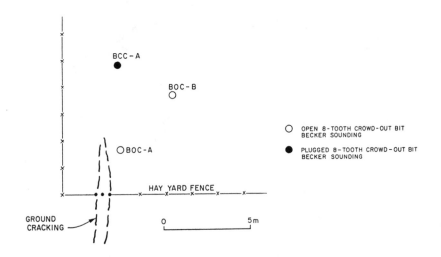

Figure 4: Plan of Pence Site A Near Hay Yard Fence

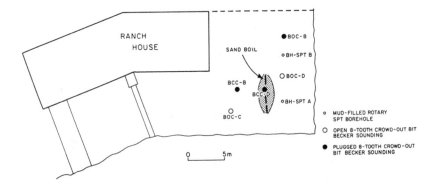

Figure 5: Plan of Pence Site B Near Ranch House

All of the Becker boreholes were drilled using a 168-mm O.D. casing. Open-bit soundings were used to obtain samples of the deposits using reverse air recirculation. Plugged-bit soundings were used to obtain penetration resistance. Penetration resistance measured with the Becker equipment was converted into equivalent SPT N_{60} blowcounts (blows per 30 cm) using the procedures developed by Harder and Seed (1986) and Harder (1988).

Gradations of Samples Obtained at Pence Ranch

Figures 6 and 7 present grain size distributions for samples obtained from open-bit Becker soundings at the two Pence Ranch Sites. These samples were obtained using the reverse air-recirculation process during driving of the Becker casing and represent highly disturbed samples of the deposits. Samples were obtained in 1 to 2-m depth increments during driving down to maximum depths of 12 m. In general, the samples consist of gap-graded silty, sandy gravels. The gravel contents for these samples ranged between 36 and 74 percent. Fines contents ranged between 1 and 11 percent. Except for samples obtained within the upper 1.5 m, the finer portions of the samples recovered were found to be non-plastic.

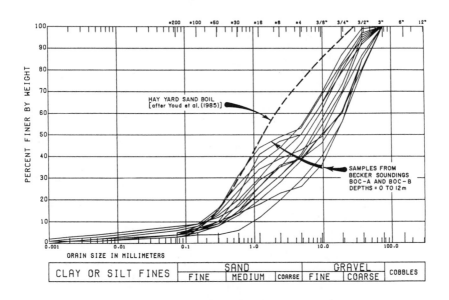

Figure 6: Gradations of Soil Samples Recovered in Becker Open-Bit Soundings Made at Pence Ranch Site A

Also shown in both figures is the gradation of the sand boil found at each site by Youd et al. (1985). The sand boil at each site has a significantly finer gradation. Presumably, the differences between the sand boil deposit and the soils in the underlying deposits are caused by the fact that the larger particles in the ground were not able to be carried to the surface by the upward flow of water resulting from the earthquake-induced pore pressures.

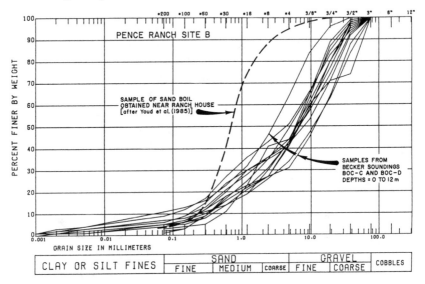

Figure 7: Gradations of Soil Samples Recovered in Becker Open-Bit Soundings Made at Pence Ranch Site B

1984 Becker Penetration Test Results at Pence Ranch Sites

Figure 8 presents the values of equivalent SPT resistance predicted by the plugged-bit Becker soundings performed at Pence Ranch Sites A and B. The values presented have been normalized to an effective overburden pressure of approximately 100 kPa - i.e. they are equivalent $(N_1)_{60}$ values. The normalization procedure was based on the following assumptions/conditions:

1. The water table was found at a 1.5 m depth.
2. The saturated density of the soils is approximately 2.08 Mg/m^3.
3. The C_N normalization curve is the one developed by Harder (1988) for use with gravelly soils having D_{50} values of approximately 5 to 20 mm and a relative density of about 50 percent.

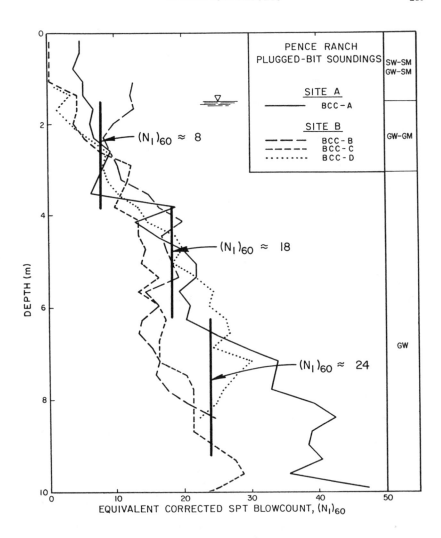

Figure 8:   Equivalent SPT Blowcounts Derived from Becker
            Tests Performed at Pence Ranch Sites A and   B

        In general, the plugged-bit sounding from Site A
gave results in good agreement with the three plugged-bit
soundings from Site B.   This suggests that the general
area has gravelly soils that are more or less uniform in
lateral extent.   This is consistent with the fact that
liquefaction and lateral spreading was observed over a

relatively large area. All four plugged-bit soundings
predict a very low normalized penetration resistance in
the upper 1.5 m of the deposit. However, because this
zone is not saturated, it could not have been responsible
for the liquefaction that developed in 1983.

In the depth interval between 1.5 and 4 m, where the
soil is saturated, the normalized equivalent SPT
resistance is approximately 8 blows per 30 cm. The
resistance in the upper saturated zone between 1.5 and
2.5 m is even lower. Although somewhat higher than that
determined for the top 1.5 m, this is still a very low
penetration resistance. At greater depths, the
normalized equivalent SPT resistance increases
significantly. In the depth interval between 4 and 6 m,
the normalized equivalent SPT blowcount is approximately
18. Between 6 and 9 m, the normalized equivalent SPT
blowcount is about 24.

It is apparent from the penetration resistance shown
in Figure 8 that the critical depth is between 1.5 and
4 m. Within this depth interval, the average gravel
content is approximately 54 percent, and the average
fines content is about 5 percent. As noted above, the
average equivalent normalized SPT blowcount, $(N_1)_{60}$, is 8
blows per 30 cm within the critical layer.

LIQUEFACTION OF GRAVELLY SOILS AT WHISKEY SPRINGS

General Description

The Whiskey Springs slide is a lateral spread that
occurred close to the epicenter of the 1983 Borah Peak
earthquake (see Figure 1). Harder (1988) employed damage
intensity values and the directivity of the fault rupture
propagation to estimate a peak ground acceleration of
0.40g at Whiskey Springs. At this location, the edges of
alluvial fans coming down the east side of the Thousand
Springs Valley meet with the marshy deposits of the
valley floor. Figure 9 presents a photograph of the
intersection of these deposits. At this juncture, the
fan has a slope of only about 5 degrees from the
horizontal. The lateral spread consisted of a
southwesterly lateral movement of blocks of sloping
ground for an accumulated maximum movement of
approximately 1 to 1.5 m. Although spreading occurred
near and along Highway 93 for over a kilometer, the
maximum movements occurred for a stretch immediately
south of Whiskey Springs.

In their reconnaissance report, Youd et al. (1985)
stated that the highway and underlying ground were

displaced at least 1.2 m laterally to the west at Whiskey Springs. Fissures as much as 1.2 m wide and 3 m deep marked the head of the lateral spread about 30 m upslope from the highway. The toe of the spread consisted of a 1.2-m high buckling of the marsh sod approximately 30 m downslope of the highway. Sand boils of silty material erupted along cracks along the road and along the toe of the spread. Figure 10 presents a cross section of the main portion of the lateral spread.

Figure 9:   Photograph Taken Along Highway 93 Approaching
             the Whiskey Springs Slide from the South
             (Photograph Taken July 1985 by L. F. Harder)

1985 Explorations at Whiskey Springs

Explorations were conducted in July 1985 at this site by Andrus et al. (1987) and others with several different types of exploration tools. The purpose of these studies was to document and characterize the soils involved in the lateral spread. Because deposits of ejected silt and fine sand were found near the fissures, the initial conclusion had been that liquefaction of a weak silty sand/sandy silt layer was responsible for the lateral movements. In fact, the responsible soil was found to be a weak silty gravel.

The explorations at this site were conducted under the direction of Professor T. L. Youd from Brigham Young University. Under his direction, 4 exploration sites were laid out on the lateral spread. Site 1 was located near the toe of the spread near the buckled sod. Site 2 was located near the upslope edge of Highway 93. Sites 3 and 4 were located further upslope, with Site 3 situated below the largest set of fissures and Site 4 situated upslope from the fissures.

At these sites, drilling explorations using hollow stem auger sampling techniques were performed. Both SPT

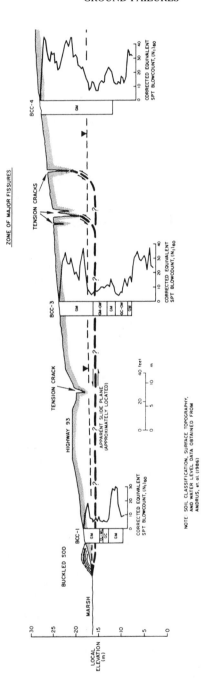

Figure 10: Cross Section of Whiskey Springs Lateral Spread Showing Corrected
Equivalent SPT Blowcounts Derived From Becker Penetration Tests

and 127-mm dry core sampling were performed through the hollow stem augers. Additional investigations consisting of CPT, trenching, cross hole and surface wave velocity testing were also performed. This information is summarized by Andrus and Youd (1987).

Becker Penetration Tests were also performed at the site in July 1985 after it became clear that the suspect layer and overlying materials were principally gravelly soils. Plugged-bit Becker soundings were performed at Sites 1, 3, and 4. In addition, an open-bit Becker sounding was performed at Site 3 approximately 20 feet away from the plugged-bit sounding performed at this site. Although performed a year after the explorations at the Pence Ranch, the same set of Becker equipment was used. Figure 11 shows a photograph of the Becker drill rig operating upslope of the highway near the conventional rotary drill rig used to perform SPT tests. As before, the procedures developed by Harder and Seed (1986) and Harder (1988) were used to convert the data into equivalent SPT blowcounts.

Figure 11:   Photograph of Becker Drill Rig (left) and Conventional Rotary Drill Rig Upslope of Highway 93 at the Whiskey Springs Slide (Photograph taken July 1985 by L. F. Harder)

1985 Becker Penetration Test Results at Whiskey Springs

Figure 12 presents the equivalent SPT blowcounts predicted by the three plugged-bit Becker soundings after being corrected to an overburden pressure of approximately 100 kPa (also shown in Figure 10). The normalization process followed the same procedure as that discussed previously for the Pence Ranch data, except

that the appropriate depth to the water surface was used
for each sounding. Also shown in Figure 12 are the
classification results determined by Andrus et al. (1987)
for the four sites explored.

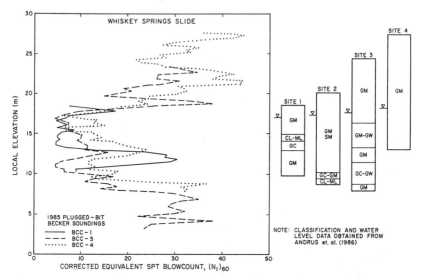

Figure 12:   Equivalent SPT Blowcounts Derived from Becker
             Tests Performed at Whiskey Springs Sites

        The results from the Becker soundings generally
define two low blowcount layers. The upper low blowcount
layer lies approximately between local elevations 13 and
17.5 m and has an average corrected equivalent SPT
blowcount, $(N_1)_{60}$, of about 8 blows per 30 cm. According
to classification tests performed by Andrus et al. (1987)
of Becker and dry core samples, this layer is composed
principally of a silty gravel. The second low blowcount
layer lies at a greater depth, approximately between
local elevations 8.5 and 11.5 m. This lower interval had
a slightly higher blowcount, was sometimes found to have
a greater percentage of clayey fines, and may not be
liquefiable. The slightly sloping ground water surface
was found near the top of the upper low blowcount layer.

        Presented in Figure 13 are gradations from samples
obtained within the critical layer using the 127-mm I.D.
dry coring sampler and the 109-mm I.D. Becker open-bit
drive bit. Although there is a range in the data, the
average gravel content is approximately 57 percent, and
the average fines content is about 20 percent. With
coefficients of uniformity ranging between 200 and 1600,
however, it is evident that the larger gravel particles

are simply floating without direct contact with each other in a silty sand matrix. Except for two samples with plasticity indexes of 1 and 2, the other samples from this layer were found to be non-plastic.

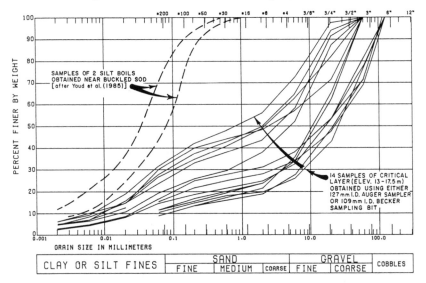

Figure 13: Gradations of Samples Recovered from the Suspect Liquefied Soil Layer at Whiskey Springs (Local Elev. 13 - 17.5 m)

Also shown in Figure 13 is the range of grain sizes developed by Youd et al. (1985) for the silt boils found at this site. As at the Pence Ranch, the larger gravel particles apparently were simply unable to be carried to the surface. This behavior has implications for other alluvial sites where only sand or silt boils on the surface are the only pieces of information available regarding the layer which liquefied. It may be that other case histories of earthquake-induced distress attributed to liquefaction of sands and/or silts were actually due to liquefaction of gravelly soils.

Residual Shear Strength of Whiskey Springs Silty Gravel

The 1.2-m lateral movement at Whiskey Springs provided an opportunity to estimate the undrained residual shear strength of the liquefied silty gravel layer. Harder (1988) used slope stability analyses, considering both post-earthquake and sliding block conditions, for this purpose. The analysis results indicated residual shear strengths between 6 and 7.5 kPa for the critical silty gravel layer at Whiskey Springs. This calculation was used as one of the case histories employed by Seed and Harder (1990) to define the relationship between SPT blowcount and residual shear strength. Although the SPT blowcount for this case history was actually derived from Becker blowcounts, the residual strength determined for this silty gravel was in good accord with residual shear strengths of sandy soils.

COMPARISONS WITH SEED ET AL. (1985) CORRELATION BETWEEN SPT BLOWCOUNT AND LIQUEFACTION TRIGGERING RESISTANCE

The application of the Becker Penetration Test relies on the development of equivalent SPT blowcounts and the use of existing correlations between behavior and SPT blowcount. Consequently, it was of interest to determine if the equivalent SPT blowcounts determined at the gravel liquefaction sites, when applied with existing correlations, could predict the actual behavior of the gravels. As a first step, the cyclic stress ratios (CSR) were computed for both the Pence Ranch and Whiskey Springs sites. The CSR values were computed using the simplified formula described by Seed et al. (1985) and Seed and Harder (1990). The CSR values were developed and modified to values comparable to level-ground cyclic stress ratios associated with Magnitude 7.5 earthquakes using the following assumptions/adjustments:

1. The mean depth for the critical layer at each site was chosen for calculation of the CSR value. The mean depth at Pence Ranch was designated as 2.5 m, and the mean depth at Whiskey Springs was designated as 7.5 m.

2. CSR values were computed using peak ground accelerations of 0.29g and 0.40g for the Pence Ranch and Whiskey Springs sites, respectively.

3. The CSR values calculated by the simplified formula were reduced by 3 percent to allow for the fact that the Borah Peak Earthquake had a magnitude that was slightly lower than the Magnitude 7.5 value used in the correlation.

4. The Pence Ranch site represented level ground conditions at relatively low overburden pressures. However, the Whiskey Springs Slide had moderate overburden pressures and a slight slope. Consequently, the calculated CSR value at Whiskey Springs was divided by a K_α value of 1.1 and a K_σ value of 0.95, as outlined by Seed

and Harder (1990) and Harder (1988) in order to obtain an equivalent level ground CSR.

In addition to the above modifications, the fines contents selected for use with the correlations represented the fines content of the sand matrix and not of the total gradation. The reason for this adjustment is the fact that existing SPT correlations are based on the performance of sandy and/or silty soils. Consequently, the effects of fine soil particles on the measured penetration and liquefaction resistance are established only for such soils. The sandy gravels which liquefied at Pence Ranch and Whiskey Springs had very large coefficients of uniformity with values generally between 20 and 1600. In such soils, the gravel particles will be floating in a sandy matrix without direct contact with each other. It would be expected that a large penetration test in the matrix material alone, existing in situ at the same void ratio as in the matrix of the total fraction, would give the same penetration resistance as would be produced by a test made in the total fraction. Thus, because existing SPT correlations are based on soils having few particles larger than 10 mm, the fines contents of the matrix were determined by the fines content of the portion finer than 10 mm.

Table 1 presents a summary of the average equivalent SPT blowcounts, level ground CSR values, and matrix fines contents for the critical soil layers which liquefied at the Pence Ranch and Whiskey Springs sites. Also shown in the table is information from Mackay Dam. As described by Harder (1992), this dumped gravel dam was also investigated using the Becker Penetration Test following the 1983 Borah Peak Earthquake. However, probably due to its larger distance from the fault rupture (see Figure 1), and corresponding lower acceleration level, the loose gravels in Mackay Dam did not fully liquefy. Thus, there are actually three sandy gravel sites which sustained shaking during the 1983 Borah Peak Earthquake, and where Becker Penetration Tests have been performed.

Presented in Figure 14 is the correlation developed by Seed et al. (1985) for predicting liquefaction potential using the results of SPT tests. Also shown on this figure are two solid symbols representing the conditions and response found at the two gravel sites, Pence Ranch and Whiskey Springs, where liquefaction was found to have occurred. In addition, there is an open symbol representing the conditions at Mackay Dam where liquefaction did not completely develop. There is excellent agreement between the plotted points and the correlation curves. Although the data is limited to only three points, the results indicate that the Becker

penetration test can be employed successfully to predict
the liquefaction potential of gravelly deposits during
earthquake shaking.

**Table 1: Summary of Results and Performance of Gravel
Sites Shaken by the 1983 Borah Peak Earthquake**

SITE	LIQUEFACTION ?	SOIL TYPE	GC (%)	SC (%)	FC (%)	MATRIX FC (%)	Equivalent SPT Blowcount, $(N_1)_{60}$	PGA (g)	Equiv. Level Ground Cyclic Stress Ratio $M_s=7.5$
PENCE RANCH	YES	GW-GM	54	41	5	8	8	0.29	0.22
WHISKEY SPRINGS	YES	GM	57	23	20	40	8	0.40	0.25
MACKAY DAM	NO	GW-GM	64	29	7	12	9	0.22	0.10

Note: GC denotes gravel content - SC denotes sand content - FC denotes fines content.
MATRIX FC denotes fines content for fraction of soil finer than 10 mm.
PGA denotes peak ground acceleration (estimated).

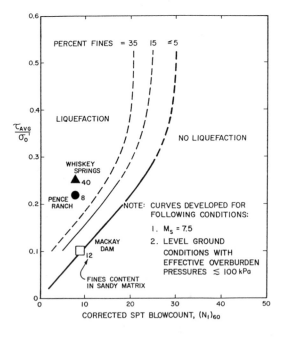

Figure 14: Comparison of Predicted and Actual
Liquefaction Performance of Gravelly Soils
During the 1983 Borah Peak Earthquake

CONCLUSIONS

The main conclusions resulting from this study may be summarized as follows:

1. Two deposits of gravelly soils, located at Pence Ranch and Whiskey Springs, liquefied during the 1983 Borah Peak Earthquake. While sand boils at both sites originally suggested that sandy and/or silty soils had liquefied, investigations have shown that sandy and silty gravels were responsible for the liquefaction-induced lateral spreads at both sites.

2. The Becker Penetration Test was an effective tool in quantifying the penetration resistance of the gravelly soils and in obtaining samples for classification testing.

3. The Becker Penetration Test, when used together with the correlations developed for sands between SPT blowcount and liquefaction potential, can be employed successfully to predict the liquefaction behavior of gravelly soils.

ACKNOWLEDGEMENTS

The two case histories of gravel liquefaction described in the proceeding pages were originally investigated by T. Leslie Youd and his associates. His assistance and coordination during the field investigations is greatly appreciated. Some of the photographs and drawings in this paper were originally presented by Dr. Youd in Earthquake Spectra, published by the Earthquake Engineering Research Institute. Their permission for reprinting is also appreciated. The studies involving the Becker Penetration Test were conducted while the author was a student under the late Dr. H. Bolton Seed. Dr. Seed's leadership, review, and guidance during the phases of this study were instrumental in their successful completion. The assistance of Ms. Laura Pence and Ms. Rita Lundy are also gratefully acknowledged.

APPENDIX.   REFERENCES

Andrus, Ronald D. and Youd, T. Leslie (1987). "Subsurface Investigation of a Liquefaction-Induced Lateral Spread - Thousand Springs Valley, Idaho," report prepared for the U. S. Army Corps of Engineers (miscellaneous paper GL-87-8), Department of Civil Engineering, Brigham Young University, Provo, Utah, May.

Crone, A., Machette, M., Bonilla, M., Lienkaemper, J., Pierce, K., Scott, W., and Bucknam, R. (1985). "Characteristics of Surface Faulting Accompanying the Borah Peak Earthquake, Central Idaho," Proceedings of Workshop XXVIII on the Borah Peak, Idaho, Earthquake, convened under the Auspices of National Earthquake Prediction and Hazard Programs, October 3-6, 1984, United States Geological Survey Open-File Report 85-290.

Harder, Jr. Leslie F. (1992). "Investigation of Mackay Dam Following the 1983 Borah Peak Earthquake," Proceedings of the Stability and Performance of Slopes and Embankments-II Specialty Conference, Berkeley, California, June 29-July 1, 1992, American Society of Civil Engineers, Geotechnical Special Publication No. 31.

Harder, Jr. Leslie F. (1988). "Use of Penetration Tests to Determine the Cyclic Load Resistance of Gravelly Soils," Dissertation submitted as partial satisfaction for the degree of Doctor of Philosophy, University of California, Berkeley.

Harder, Jr. Leslie F. and Seed, H. Bolton (1986). "Determination of Penetration Resistance For Coarse-Grained Soils Using the Becker Hammer-Drill," Report No. UCB/EERC 86/06, Earthquake Engineering Research Center, University of California, Berkeley, May.

Seed, H. Bolton, Tokimatsu, K., Harder, L. F., and Chung, Riley M. (1985). "Influence of SPT Procedures in Soil Liquefaction Resistance Evaluations," Journal of the Geotechnical Engineering Division, ASCE, Vol. 111, No. GT12, December.

Seed, Raymond B. and Harder, Jr. Leslie F. (1990). "SPT-Based Analysis of Cyclic Pore Pressure Generation and Undrained Residual Strength," Proceedings of the Memorial Symposium for H. Bolton Seed, BiTech Publications, Ltd.

Stover, Carl W. (1985). "The Borah Peak, Idaho Earthquake of October 28, 1983--Isoseismal Map and Intensity Distribution," Earthquake Spectra, Earthquake Engineering Research Institute, Vol. 2, No. 1, November.

Youd, T. L., Harp, E. L., Keefer, D. K., and Wilson, R. C. (1985). "The Borah Peak, Idaho Earthquake of October 28, 1983--Liquefaction," Earthquake Spectra, Earthquake Engineering Research Institute, Vol. 2, No. 1, November.

IN-SITU MEASUREMENTS OF DYNAMIC SOIL PROPERTIES AND LIQUEFACTION RESISTANCES OF GRAVELLY SOILS AT KEENLEYSIDE DAM

Ken K.Y. Lum[1], M.ASCE and Li Yan[2]

ABSTRACT

This paper discusses the in-situ measurements of penetration resistance and shear wave velocity, and the evaluation of liquefaction resistance for gravelly soils at the Hugh Keenleyside Dam. The field measurements include the Standard Penetration Test (SPT), the Becker Penetration Test (BPT), and shear wave velocity using crosshole, downhole and the Spectral-Analysis-of-Surface-Waves (SASW) methods. The difficulties in obtaining meaningful SPT-N_{60} from the Standard Penetration Test in gravelly soils are discussed. Two methods of interpreting Becker penetration resistance, Harder and Seed's and Sy and Campanella's, are used to derive the equivalent SPT-N_{60} values. Compared to Harder and Seed's values, the equivalent SPT-N_{60} values from Sy and Campanella's approach show more variation and perhaps better reflect the influence of coarser particles in gravelly soils. Some existing in-situ shear wave velocity correlations based on penetration resistance obtained from the Standard Penetration Test in gravelly soils are examined. Although significant scatter exists, it is found that these correlations statistically agree with the data obtained from the larger scale BPT. Two independent liquefaction assessment methods, i.e. SPT-based and shear wave velocity-based methods, are used to estimate the liquefaction resistances of the gravelly soils at the Keenleyside Dam. It is found that the shear wave velocity-based method generally gives higher liquefaction resistances than the SPT-based method within the dam fill materials.

INTRODUCTION

The Hugh Keenleyside Dam (formerly Arrow Dam) is located on the Columbia River about 30 km north of the Canada-United States border (Fig. 1). The dam was completed in 1968 to provide storage for downstream flood control and to increase power generation in the United States under the terms of the Columbia River Treaty.

---

[1]B.C. Hydro, 6911 Southpoint Drive, Burnaby, B.C., Canada V3N 4X8

[2]Klohn-Crippen Integ, 600-510 Burrard Street, Vancouver, B.C., Canada V6C 3A8

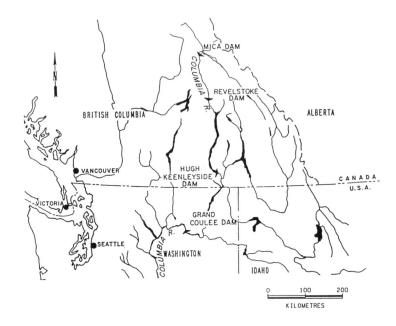

Figure 1 - Location plan

Recent seismicity studies indicate that seismic design parameters compatible with current practice are significantly higher than those used for the original design of the project over 25 years ago. Studies are currently underway by B.C. Hydro to evaluate the seismic stability of the dam with the updated seismic parameters corresponding to a Magnitude 6.5 earthquake with a horizontal peak ground acceleration of 0.22 g.

To assess the seismic stability of the dam, dynamic soil properties are required for seismic response and liquefaction analysis of the embankment. The key parameters are in-situ shear wave velocity and liquefaction resistance. The shear wave velocity is obtained either by direct in-situ shear wave measurements or by indirect empirical correlations with the Standard Penetration Test (SPT) blowcount. The liquefaction resistance in sandy soils is estimated, in practice, from existing empirical correlations with SPT blowcount or shear wave velocity. The liquefaction resistance of gravelly soils is particularly difficult to characterize because of the difficulties in obtaining undisturbed soil samples and the limitations of the SPTs in gravelly soil sites.   Because of the difficulties and uncertainties in applying the existing empirical correlations to estimate liquefaction resistances for gravelly soils, a comprehensive field investigation program was carried out at the Keenleyside Dam in an attempt to best characterize in-situ soil properties, using presently available in-situ testing techniques. This paper describes the studies undertaken to obtain penetration resistance, shear wave velocity, and liquefaction resistance data in gravelly soils at the Keenleyside Dam.

KEENLEYSIDE DAM

The Hugh Keenleyside Dam consists of an earthfill embankment about 430 m long and three concrete gravity structures (sluiceways, low level ports and a navigation lock) with a length of about 370 m. The dam is located in the Columbia Valley where the surficial soils consist of unconsolidated fluvial and glaciofluvial deposits. Although some fine-grained materials are interspersed, these deposits consist mainly of heterogeneous sands and gravels with cobbles and some boulders. The unconsolidated deposits are more than 150 m thick in the mid-valley area.

The earthfill dam is a zoned fill embankment with an upstream sloping impervious core and a downstream pervious sand and gravel shell (Figs. 2 & 3). Both upstream and downstream slopes of the dam are 3H:1V, buttressed respectively by an upstream and downstream berm. The sand and gravel portion of the earthfill dam below the original river level (about El. 419 m) was constructed by bottom dumping from barges and end dumping from trucks to form an embankment to just above the water level. The typical grain size envelope for the sand and gravel fill is shown in Fig. 4. The upstream sloping impervious core was then formed by advancing the till material from the sand and gravel embankment, and ties into the upstream blanket which extends about 670 m upstream. The remaining portion of the dam above the original river level was placed in the dry by light compaction (Golder and Bazett 1967; Bazett 1970).

FIELD TESTING PROGRAM

From 1991 to 1993, the following drilling and testing were carried out as part of the seismic investigation of the existing dam: (1) air rotary with in-situ permeability tests; (2) open bit Becker; (3) mud rotary with the Standard Penetration Test (SPT); (4) Becker Penetration Test (BPT); (5) crosshole shear wave velocity measurements; (6) downhole shear wave velocity measurements, and (7) Spectral-Analysis-of-Surface-Waves (SASW) testing.

The field test locations are shown in Fig. 2. The tests were carried out along the downstream berm, on the downstream slope and on the crest of the dam. On the upstream berm, only four underwater SASW tests were carried out. To compare the results from the different testing techniques, various methods of testing were carried out in close proximity to each other at selected locations. In this paper, only the field measurements from DH91-1, DH91-2, DH91-3 and DH93-2C located at the downstream berm and DH93-3C at the crest of the dam are discussed.

TEST RESULTS AND INTERPRETATION

Standard Penetration Tests
The SPT is the most widely used in-situ test in North America for liquefaction analysis of sandy soils using Seed's liquefaction chart (Seed et al., 1985). Although it is commonly recognized that the SPT penetration resistance N-values may be unreliable and often too high in gravelly soils due to the large gravel particle size relative to the 35 mm inside diameter of the sampler, the blowcounts obtained from the SPT have been used to evaluate the

liquefaction resistance in gravelly soils. To evaluate the gravel particle size effects on the blowcount, "small increment" blowcounts are often examined to infer the penetration resistance for the finer grained matrix of the gravelly soils (Vallee and Skryness 1979; Andrus and Youd 1987; Valera and Kaneshiro 1991).

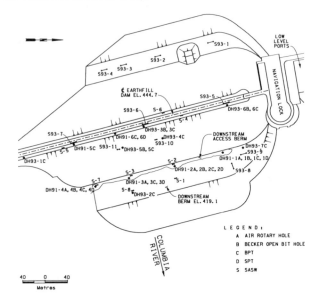

Figure 2 - Plan of earthfill dam and test locations

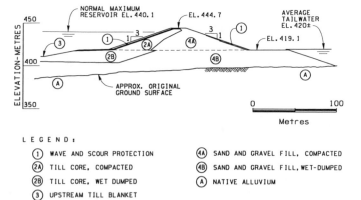

Figure 3 - Typical section of earthfill dam

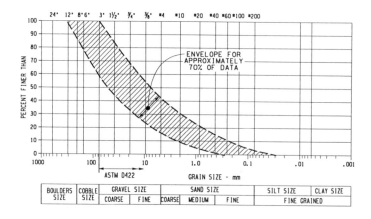

Figure 4 - Grain size envelope for Sand and Gravel fill

At the Keenleyside Dam, the SPTs were carried out with a Mayhew 1000 drill rig in drill holes advanced with a tricone bit and bentonite drilling mud. A donut hammer with the rope and cathead technique was used in the SPTs. During the test, blowcounts were recorded for every 25 mm or 50 mm penetration to provide more detailed information for evaluating the effect of gravel particles on the measured N-value.

Energy calibrations of the SPT system using force and acceleration measurements were also carried out for the drill rig. The SPT hammer energy transferred into the drill rod was calculated using the force-velocity integration method as suggested by Sy and Campanella (1991). The average transferred energy was about 41% of the theoretical hammer free-fall energy. The recorded N-values were then corrected to the standardized energy level of 60% of the theoretical free fall energy using:

$$N_{60} = N \frac{ER}{60} \tag{1}$$

where N_{60} is the SPT blowcount corrected to 60% of the hammer free-fall energy, N is the measured blowcount, and the ER is the measured energy ratio in percent and is equal to 41 in this study.

Fig. 5 shows the energy-corrected SPT blowcount, N_{60}, from the three drill holes DH91-1D, DH91-2D and DH91-3D at the downstream berm of the dam. For each hole, two sets of N_{60} values are shown. The filled circle represents the standard N_{60} value, i.e. the blowcount for a sampler penetration of 305 mm after an initial seating penetration of 150 mm. The open circle corresponds to an equivalent N_{60} estimated by multiplying the lowest blowcount for a 25 mm increment by 12 (DH91-1D and DH91-3D) or multiplying the lowest blowcount for a 50 mm increment by 6 (DH91-2D) where the blowcounts for the 25 mm increment were not available. Both individual and cumulative blowcounts of each 25 or 50 mm of penetration were also plotted to evaluate the gravel particle size effects. Suspect standard N_{60} blowcounts which may have been influenced by the gravel particles are marked on Fig. 5.

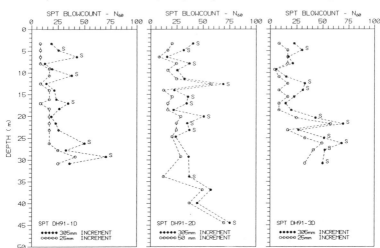

Figure 5 - SPT penetration resistances, N_{60}, for DH91-1D, DH91-2D and DH91-3D

The SPT's in the upper 20 to 25 m are within the barge- and end-dumped sand and gravel fills, whereas those at greater depths are in the foundation soils. It is seen in Fig. 5 that the SPT-N_{60} values based on the 305 mm increment are somewhat erratic, and this is not surprising. As shown in Fig. 4, the sand and gravel fill has a D_{50} ranging from 7 mm to 55 mm, and a gravel content ranging from 55% to 85%. Thus, the N_{60}-values measured from SPT with a sampler I.D. of 35 mm will inevitably be influenced by the large soil particles and gravel content, especially for the higher N-values shown in Fig. 5.

The equivalent N_{60}-values estimated from the lowest blowcount for a small increment are sometimes believed to better characterize the penetration resistance of the finer grained matrix material within the gravelly soil deposits. It is seen that these equivalent N-values are much smaller than those from the conventional 305 mm increment, particularly for the high N-values. The difference is generally smaller for the low N_{60}-values from the full penetration increment. This comparison appears to suggest that the lower N_{60}-values are less affected by the large gravel particles. Although the small increment blowcount information may provide some insight into the possible influence of large particles on the penetration resistance, the selection of the appropriate N-values for gravelly soils from the SPT is still subject to significant judgement.

Becker Penetration Tests
The BPT consists of driving a string of specially designed double-walled casings into the ground using a double acting ICE180 diesel hammer. The BPT is a larger scale penetration test that is commonly used in western North America to estimate equivalent SPT N-values in gravelly soils for foundation design and liquefaction assessment.

Typical of all diesel hammers, the Becker hammer gives variable energy output depending on variable combustion conditions and soil resistances. Thus, in order to account for the variable hammer energy output from blow to blow, Harder and Seed (1986) proposed that the measured Becker blowcounts be first corrected to a reference constant combustion condition based on hammer bounce chamber pressure measurements. Since the bounce chamber pressure is affected by atmospheric pressure, the measured bounce chamber pressure at high altitude is also corrected, if necessary, to the mean atmospheric pressure at sea level. The Harder and Seed's BPT blowcount correction procedure based on peak bounce chamber pressure is shown in Fig. 6a. With the corrected Becker blowcount, N_{bc}, the equivalent SPT-N_{60} value is then obtained from Harder and Seed's BPT-SPT correlation shown in Fig. 6b.

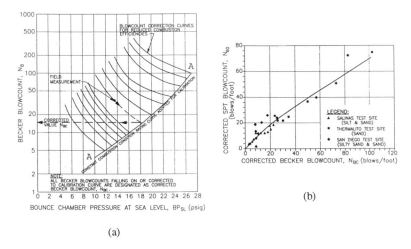

(a)

(b)

Figure 6 - (a) Harder and Seed's Becker blowcount correction chart (after Harder and Seed 1986), (b) Harder and Seed's BPT-SPT correlation (after Harder and Seed 1986).

The BPT-SPT correlation proposed by Harder and Seed, however, does not explicitly address casing friction effects in the BPT. Recent studies by Yan and Wightman (1992), Wightman et al. (1993) and Sy and Campanella (1992, 1993) have contributed to a better understanding of the effects of casing friction on the measured BPT blowcount. Sy and Campanella (1993) proposed an alternative approach to derive equivalent SPT-N_{60} values from BPT blowcounts. Their method considers the actual Becker hammer energy transferred to the Becker casing and explicitly accounts for casing friction effects on the BPT blowcounts.

In the Sy and Campanella approach, the BPT is monitored using a Pile Driving Analyzer (PDA), similar to that used in the dynamic monitoring of pile driving (ASTM D4945-89). The PDA measures force and acceleration time histories near the top of the Becker casings for each hammer blow. From the dynamic measurements, the energy transferred from the Becker hammer into the BPT casing is computed. Based on extensive energy measurements of BPTs, Sy and Campanella (1992) proposed that the measured BPT blowcount be corrected to a reference energy of 30% of the manufacturer's rated hammer energy for the diesel hammer using the following equation:

$$N_{b30} = N_b \; \frac{ENTHRU}{30} \tag{2}$$

where N_{b30} is the blowcount corrected to an energy of 30% (or 3.30 kJ), N_b is the measured blowcount, and ENTHRU is the maximum measured energy transmitted into the Becker casing expressed as percent of the rated hammer energy of 11.0 kJ for the ICE180 hammer. The energy correction for BPT blowcount in Eq. (2) is similar to the energy corrections for SPT blowcount shown in Eq. (1).

From PDA dynamic measurements, the soil friction and its distribution along the Becker casing can be estimated by a signal-matching technique of the measured force and velocity traces using the wave equation analysis program CAPWAP (Rausche et al., 1985). With the computed total shaft resistances (R_s) and the energy corrected N_{b30}, the equivalent SPT-N_{60} value can be estimated using the BPT-SPT correlation proposed by Sy and Campanella as shown in Fig. 7.

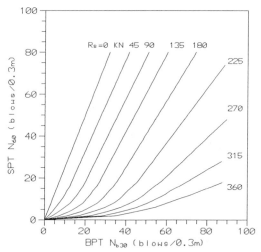

Figure 7 - BPT-SPT correlation proposed by Sy and Campanella (1993).

At the Keenleyside Dam, the BPTs were carried out through the dam fill and into the foundation soils with an AP-1000 Becker rig and 168 mm O.D. casing. During the BPT, the blowcount for each 305 mm penetration and the peak bounce chamber pressure for each hammer blow were recorded and stored in a laptop computer. Bounce chamber pressure measurements were conducted for all BPTs. PDA dynamic monitoring, however, was performed only at two BPT holes: one was located at the downstream berm (DH93-2C) and the other at the crest of the dam (DH93-3C).

Figure 8 shows the measured BPT blowcount (N_b), bounce chamber pressure (BP), and transferred energy (ENTHRU) for DH93-2C. The BP and ENTHRU values shown are the

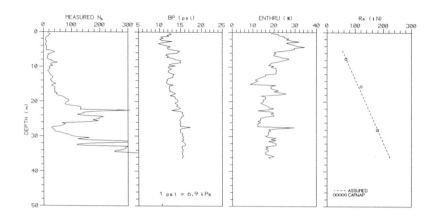

Figure 8 - Measured BPT blowcount (N_b), bounce chamber pressure (BP), and maximum transferred energy (ENTHRU) for DH93-2C.

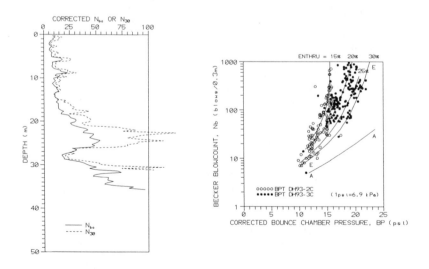

Figure 9 - Comparison of corrected Becker blowcounts, N_{bc} and N_{b30}, for DH93-2C.

Figure 10 - Contours of equal Becker energy based on data from DH93-2C and DH93-3C.

average values for each 305 mm of penetration. Also shown is the total shaft resistance (R_s) computed by the CAPWAP analysis for three selected blows at different depths. It is seen that the total shaft resistance on the Becker casing increases approximately linearly with depth. This same trend was also observed down to 50 m depth at DH93-3C.

A comparison of the bounce chamber pressure corrected blowcounts, N_{bc}, using Harder and Seed's approach and the energy corrected blowcounts, N_{b30}, following Sy and Campanella's approach, is shown in Figure 9 for DH93-2C. It is seen that the N_{b30} values are generally larger than the N_{bc} values. This would suggest that Harder and Seed's N_{bc} values are corrected to a constant combustion rating curve that corresponds to energy levels higher than 30%.

Continuous PDA monitoring of each BPT is an expensive and time-consuming process and therefore was only applied to two BPT holes in this project. Based on wave equation analyses of the BPT, Yan and Wightman (1992) and Wightman et al. (1993) showed that it is possible to construct a set of equal energy contours in the blowcount vs. bounce chamber pressure plot from field test data where measurements of both BPT transferred energies and bounce chamber pressures are available. Once this set of energy contours is established for a given rig, it allows correction of Becker blowcounts to a reference energy level for BPTs where only bounce chamber pressure measurements are available. This approximate approach for energy correction of BPT blowcount was used for the results presented below.

A site and rig specific set of Becker energy contours was established by Sy (1993) based on the bounce chamber pressure and the ENTHRU data from DH93-2C and DH93-3C. These contours of equal ENTHRU are shown in Fig. 10 along with the database used. For comparison, Harder and Seed's constant combustion (A-A) line for the AP1000 drill rig is also shown in the figure. The bounce chamber pressure-ENTHRU correlations shown in Fig. 10 are based on limited data with considerable scatter particularly at high blowcounts. Nevertheless, for BPTs performed with only bounce chamber pressure measurements and no energy data, the best fit 30% ENTHRU contour, or E-E line, appears to provide a reasonable estimate of equivalent N_{b30} values at Keenleyside Dam for blowcounts less than about 50, as shown in Fig. 11. It is also interesting to note in Fig. 10 that Harder and Seed's constant combustion (A-A) line plots much lower than the 30% (E-E) energy line, again suggesting that the A-A line represents a much higher energy condition than 30%, and, hence, requires larger corrections to the measured blowcounts.

Although Harder and Seed's approach does not explicitly address the effects of friction, their BPT-SPT correlation obtained from sand sites likely contains some friction component. This can be examined by converting those N_{bc} values in their BPT-SPT correlation (Fig. 6b) to equivalent N_{b30} values. The "correction" of N_{bc} on the A-A line to equivalent N_{b30} on the E-E line in Fig. 10 is accomplished by a procedure similar to that shown in Fig. 6a. The converted Harder and Seed's correlation is then superimposed onto the chart proposed by Sy and Campanella (1993) as shown in Fig. 12. Note that Harder and Seed's correlation cuts through Sy and Campanella's constant casing friction lines, suggesting that embedded friction in Harder and Seed's BPT-SPT correlation increases with increasing blowcount and can have a significant effect on the estimated equivalent SPT-N_{60} values.

Figure 13 shows a comparison of the equivalent SPT-N_{60} value derived from the BPT data by the two methods. The equivalent N_{60}-values from Sy and Campanella's correlations are generally higher than those obtained from Harder and Seed's correlation for moderate to high blowcounts. However, for low blowcounts at depths in excess of 15 m where casing friction is significant, Sy and Campanella's equivalent N_{60}-values are less than Harder and Seed's values. It is also interesting to note that the equivalent N_{60}-values from Sy and Campanella's method show much larger variations than Harder and Seed's equivalent SPT values. This is in part due to the unquantified effects of built-in casing friction in Harder and Seed's method, and the smaller corrections required to correct the measured BPT blowcount to the E-E line, as compared to that required to correct to Harder and Seed's A-A line.

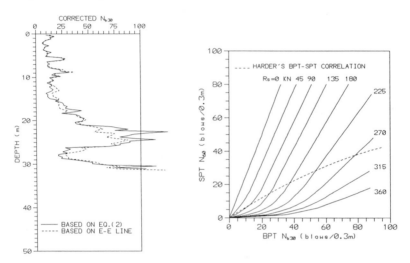

Figure 11 - Comparison of energy-corrected blowcounts, N_{b30}, using Eq. (2) and E-E line.

Figure 12 - Comparison of BPT-SPT correlations from Harder and Seed (1986) and Sy and Campanella (1993)

Shear Wave Velocity
In-situ shear wave velocity measurements were performed to obtain small strain shear modulus parameters for use in ground response analyses and as an independent field index for evaluating liquefaction resistance. Measurements were obtained using the crosshole, downhole and SASW seismic methods.

The crosshole shear wave velocity was measured at four locations in three-hole and two-hole arrays at the downstream berm. The holes were generally spaced at a distance of 4 to 6 m apart, and shear wave velocity measurements were obtained at 0.75 m depth intervals.

Downhole shear wave velocity measurements were obtained at two locations; one at the downstream berm and the other at the crest of the dam. Measurements were also made at vertical intervals of 0.75 m. Both crosshole and downhole seismic measurements were carried out inside the PVC casing installed following completion of the air rotary and Becker drill holes.

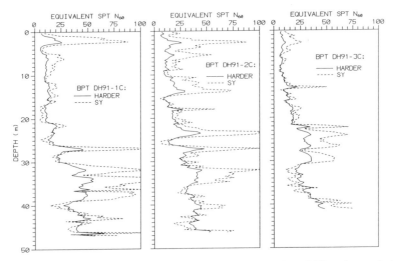

Figure 13 -   Comparison of the equivalent SPT-N_{60} values from BPTs using methods proposed by Harder and Seed (1986) and Sy and Campanella (1993).

Shear wave velocity measurements were also carried out using the non-intrusive SASW method adjacent to the crosshole sites to provide a comparison of the results, and at other locations to provide supplemental shear wave velocity data. The SASW method is based upon generation and measurement of surface Rayleigh waves. Surface waves containing different frequencies are generated by a vertical impact on the ground surface. In a vertically nonhomogeneous system, the surface waves are dispersive and the penetration depth of the surface waves increases with increasing wavelength. Thus, waves with short wavelengths propagate in the top layers and hence their velocities correspond to the properties of these shallow layers. On the other hand, waves with longer wavelengths penetrate deeper and their velocities are then influenced by the properties of the soil at depth. By forward modelling, the shear wave velocity profile can be obtained.

The measured shear wave velocity, V_s, at three locations at the downstream berm are shown in Fig. 14. The shear wave velocities obtained at these locations are reasonably similar for the crosshole, downhole and SASW methods. However, at a few other locations at the crest and downstream slope of the dam the SASW gave results which do not appear as reliable. This may be due to the sloping ground geometry, physical site and other constraints. Depending on the size of impact hammer used, the SASW appears to be limited to a penetration depth of 15 to 30 m.

In practice, when the in-situ shear wave velocity is not measured, shear wave velocity is often estimated from the SPT N-values. Various empirical correlations exist relating shear wave velocity in gravelly soils to SPT N-values (Sykora and Koester 1988). Table 1 lists some of these correlations for gravelly soils.

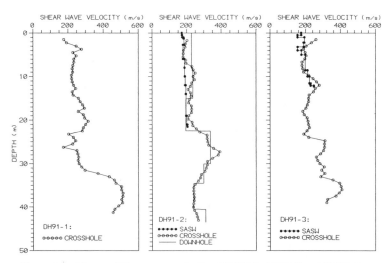

Figure 14 - Measured Shear wave velocity at DH91-1, DH91-2 and DH91-3

These correlations are based on the SPT carried out in gravelly soils. Therefore, with the shear wave velocity and the equivalent SPT-N_{60} data gathered at the Keenleyside Dam, it is worthwhile to examine these correlations against the equivalent SPT-N_{60} data obtained from a larger penetration tool such as the BPT.

Table 1 - Correlations between SPT-N and Shear Wave Velocity for gravelly soils

References	Shear Wave Velocity, V_s, (m/s)	Database	SPT energy ratio, ER*
Ohta & Goto (1978)	$V_s = 94.2 \ N_{67}^{0.34}$ (gravel)	Field data	Japanese rope and cathead, donut hammer, ER = 67%
Imai & Tonouchi (1982)	$V_s = 75.4 \ N_{67}^{0.351}$ (alluvial gravel)	Field data	Japanese rope and cathead, donut hammer, ER = 67%
Yoshida et al. (1988)	$V_s = 125 \ N_{78}^{0.25} \ (\sigma_v'/Pa)^{0.14}$ (75% gravel content)	Lab data	Japanese Tombi, donut hammer, ER = 78%

Note: * Estimated based on Seed et al. (1985)

Figure 15 plots the shear wave velocities measured from the crosshole method and the equivalent SPT-N_{60} data from the BPT using Harder and Seed's approach. A least square regression analysis was performed on the data set, and a best fit equation was found as follows:

$$V_s = 116 \ N_{60}^{0.274} \qquad (with \quad r = 0.619) \qquad (3)$$

where V_s is shear wave velocity in m/s, N_{60} is the equivalent SPT blowcount, and r is the regression correlation coefficient. This equation is shown in Fig. 15 together with the other equations listed in Table 1 and the data set used. The equation proposed by Yoshida et al. is plotted with a vertical stress level of 200 kPa which is about the average stress in the data set obtained for the Keenleyside Dam.

It can be seen that the existing correlations for gravelly soils compare reasonably well with the best fit equation, although the correlation proposed by Imai and Tonouchi gives lower shear wave velocity values. It should also be noted that significant scatter exists in the data shown in Fig. 15. Similar comparison with the equivalent SPT-N_{60} values from Sy and Campanella's approach shows a wider data spread or scatter in the horizontal axis. Therefore, the existing N_{60} vs. V_s correlations for gravelly soils appear to be statistically consistent with the data obtained from the larger scale BPT, but significant errors would still be expected when obtaining shear wave velocity from SPT or BPT penetration resistances.

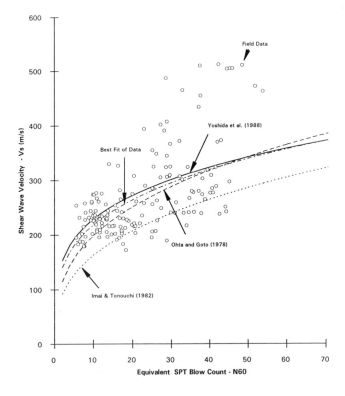

Figure 15 - Correlation between shear wave velocity and equivalent SPT-N_{60} values.

LIQUEFACTION RESISTANCE

There is increasing concern for seismic liquefaction in gravelly soils as evidenced by some recent case histories (Valera and Kaneshiro 1991). However, the present database of observed liquefaction in gravelly soils is still very limited. Current assessment of liquefaction potential in gravelly soils is mainly based on the field performance of sandy soils in past earthquakes. Two independent databases compiled from past earthquakes, one based on SPT and the other based on shear wave velocity, are used in this study to assess the liquefaction resistance.

The first approach is the SPT-based method proposed by Seed et al. (1985) which relates soil liquefaction resistance to SPT blowcount, $(N_1)_{60}$, normalized by the SPT energy and overburden pressure. Figure 16 shows Seed's liquefaction chart for M7.5 earthquakes and for different fines contents. Andrus and Youd (1987) showed that Seed's liquefaction chart can be used to predict the liquefaction observed at gravelly soil sites during the 1983 M7.3 Borah Peak earthquake, using penetration resistance obtained from the SPT. More commonly, however, BPTs are carried out in gravelly soils to estimate the equivalent SPT-$(N_1)_{60}$ values, and then to assess the liquefaction potential using Seed's liquefaction chart.

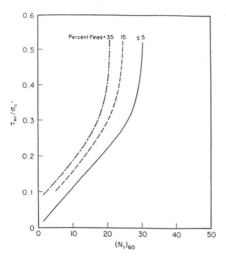

Figure 16 - Seed's SPT-based liquefaction resistance chart for M7.5 (after Seed et al. 1985).

The second approach relates the liquefaction resistance with the shear wave velocity. The shear wave-based liquefaction resistance chart proposed by Robertson et al. (1992) relates the liquefaction resistance with the normalized shear wave velocity, V_{s1}, which is obtained from:

$$V_{s1} = V_s \left(\frac{P_a}{\sigma'_{vo}}\right)^{0.25} \tag{4}$$

where V_s is the shear wave velocity, P_a is the atmospheric pressure, σ'_{vo} is the in-situ effective overburden stress in the same units as P_a. Kayen et al. (1992) subsequently updated Robertson et al.'s liquefaction assessment chart with the 1989 M7.1 Loma Prieta earthquake data, and proposed a liquefaction demarcation curve that predicts lower liquefaction resistances.

To examine the applicability of different liquefaction curves to gravelly soils, the shear wave velocity data and the estimated cyclic stresses due to the Borah Peak earthquake at the liquefied and non-liquefied gravelly soil sites (Stokoe et al. 1988; Andrus et al. 1992) were reviewed and plotted in Fig. 17 against the liquefaction resistance curves proposed by Robertson et al. and Kayen et al. Also shown in Fig. 17 are two boundary curves converted from Seed's SPT-based liquefaction curve for clean sand using different SPT-N vs. V_s correlations. The SPT-N vs. V_s correlations proposed by Seed et al. (1983) and Ohta and Goto (1978) for sand were used. These two boundary curves are plotted at an overburden pressure of 100 kPa, and, as such, are only intended to show the relative location of Seed's boundary curve in the shear wave-based liquefaction chart. It can be seen that the data from the Borah Peak earthquake appear to support the curve suggested by Kayen et al. Thus, Kayen et al.'s curve is used in the present analysis. It is also worth noting that Seed's boundary curve is located to the right of Kayen et al.'s curve, possibly indicating that Seed's SPT-based approach gives lower liquefaction resistance than Kayen et al.'s curve for overburden pressures of about 100 kPa.

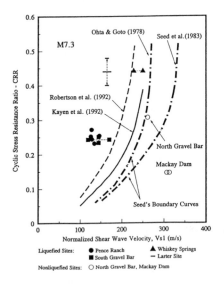

Figure 17 - Shear wave-based liquefaction resistance chart (modified after Kayen et al. 1992).

The estimated liquefaction resistances for M6.5 earthquake at DH91-1, DH91-2 and DH91-3 using the SPT- and the shear wave-based methods are shown in Fig. 18. For the SPT-based method, equivalent N_{60}-values from both BPT methods and measured SPT-N_{60} from the standard 305 mm increment were used. It is seen that the liquefaction resistance estimated

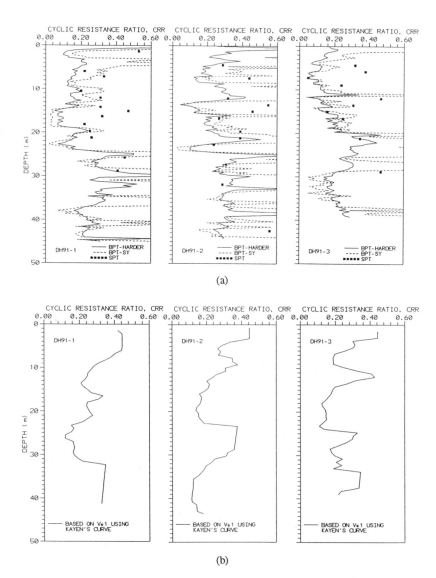

Figure 18 - Estimated liquefaction resistances for M6.5 for DH91-1, DH91-2 and DH91-3 using (a) SPT-based method and (b) shear wave-based method.

from Sy and Campanella's BPT data shows more variation, and, at some locations at depth, gives lower resistance than Harder and Seed's. The liquefaction resistances from the SPT also show considerable variations. However, it is also seen that the lower resistance values from the SPT compare reasonably close with those from the BPT, suggesting that these lower SPT blowcounts may be less affected by the gravel particles. Comparison of the liquefaction resistances from SPT and shear wave velocity indicates that the shear wave-based method generally gives higher resistance than that from the SPT-based method within the dam fill. In the foundation soils, however, where the soil conditions are more variable, the trend is not consistent.

SUMMARY AND CONCLUSIONS

The paper presents the in-situ measurements of penetration resistance and shear wave velocity, and the evaluation of liquefaction resistance for gravelly soils at the Keenleyside Dam. Difficulties in obtaining meaningful SPT-N_{60} from the SPT in gravelly soils are discussed. Two methods of interpretating Becker penetration resistance are used to derive the equivalent SPT-N_{60} values. The method proposed by Harder and Seed (1986) has been widely used in North America, but it does not explicitly address casing friction. A more recent method proposed by Sy and Campanella (1993) accounts for the variation of Becker hammer energy and casing friction effects in the BPT. The equivalent SPT-N_{60} values derived using Sy and Campanella's approach show more variation and is more sensitive than those using Harder and Seed's approach.

Crosshole, downhole, and SASW shear wave velocity measurements were carried out at the Keenleyside Dam, and are shown to give comparable results at the downstream berm locations. At a few other locations, the SASW gave results which do not appear as reliable. Some existing in-situ shear wave velocity correlations based on penetration resistance obtained from the SPTs in gravelly soils are examined. It is found that correlations based on SPT resistance are statistically consistent with the data obtained from the BPT at this site, but significant errors would still be expected in obtaining equivalent shear wave velocity from penetration resistance.

Two independent liquefaction assessment methods, SPT-based and shear wave-based methods, are used to assess the liquefaction resistance of gravelly soils at the Keenleyside Dam. The method based on shear wave velocity is reviewed in the light of reported field data from gravelly soil sites during the Borah Peak earthquake. The available limited database appears to support the liquefaction curve proposed by Kayen et al. (1992) based on Loma Prieta earthquake data. Comparison of the liquefaction resistances from penetration data and shear wave velocity data indicates that the shear wave-based method generally gives higher resistance than that from the SPT-based method within the dam fill at the Keenleyside Dam.

ACKNOWLEDGEMENTS

The authors wish to acknowledge B.C. Hydro and Power Authority for permission to publish this paper, and Dr. Alex Sy for his helpful suggestions and input in this project with regards to interpretation of the Becker Penetration Tests. The opinions expressed in the paper are those of the authors and do not necessarily represent those of any other individual or organization.

APPENDIX - REFERENCES

Andrus, R.D. and Youd, T.L. 1987. Subsurface investigations of a liquefaction-induced lateral
   spread. Thousand Springs Valley, Idaho.   U.S. Army Corps of Engineers,
   Geotechnical Laboratory Miscellaneous Paper GL-87-8, 131 p.
Andrus, R.D., Stokoe, K.H. , Bay, J.A. and Youd, T.L. 1992. In situ V_s of gravelly soils
   which liquefied. Proc. 10th World Conference on Earthquake Engineering, Madrid,
   Spain, 1447-1452.
ASTM Standard D4945-89. Standard test method for high-strain dynamic testing of piles,
   ASTM Standards, Vol. 04.08, 1018-1024.
Bazett, D.J. 1970. The Characteristics of till placed under water at the Arrow Dam. Proc. 10th
   Intl. Congress on Large Dams, Montreal, 1:805-821.
Golder, H.Q. and Bazett, D.J. 1967. An earth dam built by dumping through water.
   Proceedings, 9th International Congress on Large Dams, Istanbul, Turkey, 369-387.
Harder, L.F. Jr. and Seed, H.B. 1986. Determination of penetration resistance for coarse-
   grained soils using the Becker hammer drill.   Report No. UCB/EERC-86/06,
   University of California, Berkeley, USA, 118 p.
Imai, T. and Tonouchi, K. 1982. Correlation of N value with S-wave velocity and shear
   modulus. Proc. 2nd European Symp. on Penetration Testing, Amsterdam, 67-72.
Kayen, R.E., Mitchell, J.K., Seed, R.B., Lodge, A., Nishio, S. and Coutinho, R. 1992.
   Evaluation of SPT-, CPT-, and shear wave-based methods for liquefaction potential
   assessment using Loma Prieta data. Proc. 4th US-Japan Workshop on Earthquake
   Resistant Design of Lifeline Facilities and Counter-measures for Soil Liquefaction,
   NCEER-92-0019, Ed. M. Hamada & T.D.O'Rourke, 1:177-204.
Ohta, Y. and Goto, N. 1978. Empirical shear wave velocity equations in terms of
   characteristic soil indexes. Earthquake Engineering and Structural Dynamics, 6:167-
   187.
Rausche, F., Goble, G.G. and Likins, G.E. Jr. 1985. Dynamic determination of pile capacity.
   ASCE Journal of Geotechnical Engineering, 111(3):367-383.
Robertson, P.K., Woeller, D.J. and Finn, W.D.L. 1992. Seismic cone penetration test for
   evaluating liquefaction potential under cyclic loading. Canadian Geotechnical Journal,
   29(4):686-695.
Seed, H.B., Idriss, I.M., and Arango, I. 1983. Evaluation of liquefaction potential using field
   performance data. ASCE Journal of Geotechnical Engineering, 109(3):458-482.
Seed, H.B., Tokimatsu, H., Harder, L.F., and Chung, R.M. 1985. Influence of SPT procedures
   in soil liquefaction resistance evaluations. ASCE Journal of Geotechnical Engineering,
   111(12):1425-1445.
Sy, A. and Campanella, R.G. 1991. An alternative method of measuring SPT energy. Second
   International Conference on Recent Advances in Geotechnical Earthquake Engineering
   and Soil Dynamics, March 11-15, St. Louis, Missouri, 1:499-505.
Sy, A. and Campanella, R.G. 1992. Dynamic performance of the Becker hammer drill and
   penetration test. 45th Canadian Geotechnical Conference, Toronto, Ontario, Paper 24:
   1-10.
Sy, A. 1993. Dynamic measurements and interpretations of BPT at Keenleyside Dam,
   Castlegar, B.C. Report to B.C. Hydro, June 8, 24p.
Sy, A. and Campanella, R.G. 1993. BPT-SPT correlations with consideration of casing
   friction. 46th Canadian Geotechnical Conference, Saskatoon, Saskatchewan, 401-411.
Stokoe, K.H.II, Andrus, R.D., Rix, G.J., Sanchez-Salinero, I. ,Sheu, J.C. and Mok, Y.J. 1988.
   Field investigation of gravelly Soils which did and did not liquefy during the 1983

Borah Peak, Idaho, earthquake. Geotechnical Engineering Report GR87-1, Civil Engineering Department, The University of Texas at Austin.

Sykora, D.W. and Koester, J.P. 1988. Correlation between dynamic shear resistance and standard penetration resistance in Soils. Proc. of Earthquake Engineering and Soil Dynamics II, Recent Advances in Ground-Motion Evaluation, ASCE specialty conference, Utah, Ed. J.L.VonThun, 389-404.

Valera, J.E. and Kaneshiro, J.Y. 1991. Liquefaction analysis of rubber dam and review of case histories of liquefaction of gravels. Proc. 2nd Intl. Conf. on Recent Advances in Geotechnical Earthquake Engineering and Soil Dynamics, St. Louis, Missouri, 1:347-356.

Vallee, R.P. and Skryness, R.S. 1979. Sampling and in-situ density of a saturated gravel deposit. Geotechnical Testing Journal, ASTM, 2(3):136-142.

Wightman, A., Yan, L. and Diggle, D. 1993. Improvements to the Becker penetration test for estimation of SPT resistance in gravelly soils. 46th Canadian Geotechnical Conference, Saskatoon, Saskatchewan, 379-388.

Yan, L. and Wightman, A. 1992. A testing technique for earthquake liquefaction prediction in gravelly soils. Industrial Research Assistance Program (IRAP) Report by Foundex Exploration Ltd. and Klohn Leonoff Ltd. to National Research Council of Canada, Contract No. IRAP-M 40401W, 228 p.

Yoshida, Y., Ikemi, M. and Kokusho, T. 1988. Empirical formulas of SPT blowcounts for gravelly soils, Proc. of the first international symposium on penetration testing, Penetration Testing 1988, Orlando, 381-387.

A PRACTICAL PERSPECTIVE ON LIQUEFACTION OF GRAVELS

Julio E. Valera[1], M., Michael L. Traubenik[2], M.,

John A. Egan[2], M., and Jon Y. Kaneshiro[3], M.

ABSTRACT

In recent years it has become standard practice to thoroughly analyze the potential for liquefaction of gravels for critical facilities, whereas in the past this hazard was more easily dismissed based on gradation and permeability considerations alone. During this period the authors have been involved in several detailed analyses of sites underlain by gravelly soil deposits. As part of these studies, case histories cited in the literature as justification that gravel deposits can indeed liquefy were critically reviewed. The findings of this review of case histories are presented herein. The potential for liquefaction at a major facility site underlain by gravelly deposits, and relatively close to the 1989 Loma Prieta earthquake rupture, was evaluated by the authors for design-level earthquake ground motions, as well as the 1989 Loma Prieta earthquake. The results of analyses using simplified empirical procedures indicate that localized liquefaction of the gravelly soils should have occurred as a result of the 1989 event, whereas no evidence of such occurrence was observed.

INTRODUCTION

A geotechnical study for a new major expansion of a water treatment facility in Central California revealed that the site was underlain by sand and gravel deposits susceptible to liquefaction during strong earthquake ground shaking. The liquefaction potential of the sands could be evaluated using commonly accepted empirical approaches based on the Standard Penetration Test (SPT) (Seed, et al., 1985). Evaluating the liquefaction potential of the gravelly deposits was, however, not as straightforward because measured SPT blow counts can be affected by the gravel particles.

Some investigators (Harder, 1988) have proposed using a Becker Hammer Penetration Test (BHPT) to evaluate the liquefaction potential of gravelly soil deposits. Using the Becker Hammer to establish "equivalent" SPT blow counts is possible; however, costs associated with mobilizing a Becker Hammer rig and performing BHPTs are quite

[1]ESA Consultants Inc., Mountain View, California; [2]Geomatrix Consultants Inc., San Francisco, California; [3]Engineering Sciences Inc., La Jolla, California

expensive when compared to more conventional exploration techniques. These costs are often difficult to justify for standard geotechnical studies, especially given that documented cases of the liquefaction of saturated gravelly soil deposits are quite rare in the literature.

This rarity of case histories may be due to the fact that, in nature, gravelly deposits are generally free-draining, which enables partial, if not full, dissipation of excess pore pressures during earthquake ground shaking, and inhibits high excess pore pressure accumulation that could lead to liquefaction behavior. However, under certain conditions in which the materials are in a very loose state, contain large quantities of finer soil constituents, have impeded drainage, or experience extended duration of strong ground shaking, it is conceivable that gravelly deposits may liquefy (or develop high excess pore pressures) during strong earthquake ground shaking. Developing an understanding of the conditions contributing to gravelly soil deposit liquefaction was an objective of the review of reported gravelly soil liquefaction case histories, so as to translate that information into practice for applications to projects such as the site discussed herein.

REVIEW OF CASE HISTORIES

In recent years, several cases in which earthquake-induced liquefaction of gravelly deposits have reportedly occurred have been described in the literature (Harder, 1988, Hynes-Griffin, 1988). These case histories, herein referred to as Cases 1 through 7, include the following:

Case 1: Liquefaction of a gravelly-sand alluvial fan deposit during the 1948 Fukui, Japan, earthquake.

Cases 2 and 3: A flow slide in a liquefied alluvial fan containing large zones of gravelly sand and sandy gravel at Valdez and bridge foundation behavior during the 1964 Alaska earthquake.

Case 4: A slide in the upstream gravelly-sand shell of Shimen Dam during the 1975 Hanking, China, earthquake.

Case 5: A slide in the upstream sandy-gravel slope protection layer of Baihe Dam during the 1976 Tangshan, China, earthquake.

Cases 6 and 7: Liquefaction of gravelly soils in level ground at the Pence Ranch and in sloping ground at Whiskey Springs during the 1983 Borah Peak, Idaho, earthquake.

Detailed review of these case histories was carried out to gain a better understanding of the various factors and conditions responsible for the observed liquefaction effects. The findings of this review are summarized herein for each case.

Case 1 - Alluvial Fan, 1948 Fukui, Japan, Earthquake

This extensive paper by Professor Ishihara (1985) contains only a very brief statement regarding the reported liquefaction of gravelly soils in an alluvial fan deposit during the 1948 M7.3 Fukui, Japan, earthquake:

"Occasionally, however, cases are reported where liquefaction apparently occurred in gravelly soils. For instance, at the time of the Fukui earthquake of June 28, 1948 in Japan, signs of disastrous liquefaction were reportedly observed in a gravelly sand in an area of fan deposit near the epicenter of the earthquake."

No further documentation is provided by Ishihara (1985) regarding this case history by which one can evaluate the observed liquefaction effects. Subsequently, a study of liquefaction-related damage and ground failure observations/data for various locations throughout the Fukui (alluvial) Plain, as well as detailed soil conditions in Morita-cho, an area of the Plain with particularly severe liquefaction and related damage, was performed and presented by Hamada et al. (1992). The stratigraphic profiles presented in that paper contain substantial deposits of medium-dense to dense sand with gravel and sandy gravel. Hamada et al. (1992) reached the interpretation that these gravelly deposits experienced liquefaction during the 1948 Fukui earthquake, but temper that interpretation with the following:

"The shallow sandy gravel layers liquefied in spite of the high N-values measured in standard penetration tests because interbedded sand and the sand matrix were considered to be loose."

This comment by Hamada et al. (1992) underscores the greater role of the finer-grained matrix, than that of the gravel, in controlling the liquefaction behavior of this gravelly deposit.

Case 2 - Valdez, 1964 Alaska Earthquake

A report by Coulter and Migliaccio (1966) describing effects at Valdez associated with the 1964 $M_w 9.2$ (M_w = moment magnitude) Alaska earthquake rarely mentions observations of liquefaction of gravelly soils. During the earthquake, a large submarine landslide occurred within the delta at the Port of Valdez that encompassed and destroyed an extensive portion of the Valdez waterfront. Subsurface conditions at Valdez are described in the report as a thick section of deltaic deposits, composed of a poorly consolidated silt, fine sand, and gravel. The silt and fine sand occur as beds and stringers and also are widely disseminated throughout the coarser fractions. Three exploratory borings drilled on the Valdez waterfront to depths as great as 40 m reportedly showed remarkable horizontal continuity and vertical uniformity within the two distinct layers encountered. The surface layer, which was 6 to 9 m thick, was a loose to medium-dense sandy gravel with cobbles and silt. Local residents and old photographs indicate that this material was fill placed during development of the harbor facilities. The gravel layer was underlain by loose to medium-dense gravelly sand containing thin lenses of silt that extended to the maximum depths drilled.

Descriptions of ground breakage presented by Coulter and Migliaccio (1966), including development of fissures and sand boils during the earthquake, include the following:

*"During the earthquake and while the slide was taking place along the waterfront, an
extensive system of fissures developed across the Valdez delta. Some of these fissures
reportedly were opening and closing during the tremors, and considerable volume of
water and suspended silt and sand were pumped from many of them.... The largest
individual longitudinal-fissure segment observed was located 1500 feet east of Dike
Road and 800 feet north of Richardson Highway. A large volume of fine sand and silt
was ejected from this fissure.... Large volumes of ejected silt, sand and in some places
pebbles characterize the central part of the longitudinal complex."*

This last excerpt is the only observation in the entire Coulter and Migliaccio (1966)
report mentioning that liquefaction of coarse granular materials may have occurred in
the Valdez vicinity during the earthquake.

Based on a careful review of the report, direct evidence to support statements made by
other investigators that earthquake-induced liquefaction of gravelly soils occurred could
not be documented. However, indirect evidence does suggest that liquefaction of
surficial loose to medium-dense sandy gravel deposits (probably loosely dumped fill
material) at the Port of Valdez probably occurred during the earthquake.

Coulter and Migliaccio (1966) also report observations of the behavior of the Mineral
Creek fan, an area underlain by coarse alluvial gravel and a site identified for relocating
the town of Valdez. The report states the following:

*"The absence of evidence of ground breakage on the Mineral Creek fan indicates that
the coarse subsoils at the relocation site react favorably under seismic conditions...
subsurface data show that the Mineral Creek alluvial fan is underlain by more than 100
feet of medium dense to very dense cobble gravel in a matrix of medium to coarse sand."*

This excerpt clearly shows that these medium-dense to dense gravelly deposits behaved
quite well during the earthquake; evidence of liquefaction or lateral spreading was not
observed.

Case 3 - Bridge Foundation Performance, 1964 Alaska Earthquake

Extensive damage to highway bridges was caused by the 1964 Alaska earthquake. A
paper by Ross et al. (1969) documents observed bridge damage that occurred along
several highways in Alaska. The following excerpts from that report appear to describe
liquefaction-related damage:

*"The greatest concentration of bridges that sustained severe foundation movements
were founded on pilings driven through saturated sands and silts of low-to-medium
relative density (measured blow counts less than 20).... Bridge foundations that were
founded in gravels and gravelly sands (regardless of N-values) rather than sands and
silts behaved relatively well, with a generally even distribution between "nil" and
"moderate" foundation displacement, and one or two cases of moderate-to-severe
displacement."*

Based on these excerpts, it could be interpreted that the gravelly soil deposits upon which
the bridges were founded did not liquefy or did not experience significant liquefaction-
related effects. The grain size distribution range for samples obtained at the bridge sites

where the subsoil consisted of gravels and gravelly sands and where foundation displacements were minor to moderate, is shown on Figure 1.

A recent study by Bartlett and Youd (1992) closely examined the liquefaction-related ground failure phenomena experienced at a number of these bridge locations. Their study included detailed assessment of subsurface conditions and soil characteristics, as well as evaluations of the extent of liquefaction that occurred at the sites. Bartlett and Youd (1992) have interpreted, based on simplified empirical liquefaction procedures and analysis of lateral spreading displacements, that the zones of liquefaction at various bridges included strata of sandy gravel and/or gravelly sand. The grain size characteristics of these strata are consistent with the range presented by Ross et al. (1969) shown on Figure 1. Generally, the gravelly deposits interpreted to have liquefied were loose to medium dense and commonly had SPT blow count values of 20 or less. The following comments made by Bartlet and Youd (1992) may help explain the better bridge foundation performance cited by Ross et al. (1969):

"Particle size also influences liquefaction susceptibility and displacement. Permeable soils dissipate excess pore pressures more rapidly, thus preventing or reducing the time soil is liquefied and subject to mobilization. For example, liquefied sandy gravel and fine gravel do not appear to displace as much as clean sand."

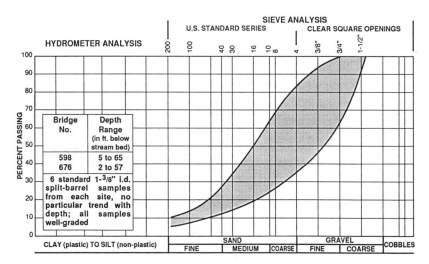

Figure 1. Grain-size distributions of Foundation Soils at Alaskan Bridge Sites
(after Ross et. al., 1969)

Case 4 - Shimen Dam, 1975 Hanking, China, Earthquake

Shortly after the occurrence 1975 M7.3 Hanking, China, earthquake, a slide occurred in the saturated upstream shell of the Shimen Dam. Modified Mercalli intensities at the dam, which was located a distance of approximately 33 km from the earthquake epicenter, were on the order of VII. The upstream shell of the dam consisted of a loosely placed well-graded sand-gravel mixture (Figure 2). The slide was very shallow having a maximum depth of about 4.6 m (infinite slope failure-type) and occurred some 80 minutes after the earthquake. Wang (1984) attributed failure to development of high excess pore pressures or liquefaction of the surficial, relatively loose (and not-very-pervious) sand-gravel shell.

Case 5 - Baihe Dam, 1976 Tangsham, China, Earthquake

During the 1976 M7.8 Tangshan, China, earthquake, a flow slide occurred in the saturated upstream protection zone of Baihe Dam. Peak ground horizontal accelerations recorded at the site near the dam crest were on the order of 0.15 g; accelerations of about 0.05 g were recorded near the downstream toe. The upstream protection zone of the dam overlies the sloping clay core and consists of a gap-graded sand-gravel mixture containing up to 60 percent sand and fines and averaging about 38 percent sand and fines (Figure 2). This slide was also of the infinite slope failure-type and extended over the entire length of the dam. Lateral spreading occurred within the reservoir as far as 100 m from the upstream toe of the dam.

No information is provided on the method or degree of compaction of the sand-gravel protection zone. However, a relative density of 56 percent, dry density of 1.65 g/cc, and a permeability of 10^{-3} cm/sec have been assigned to the fine-grained portion of the material (Wang, 1984). These data indicate that the materials are generally loose and not

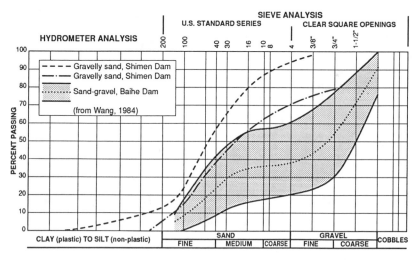

Figure 2. Gradation of Soils Which Liquefied During Past Earthquakes in China

very free-draining. Wang (1984) states that segregation between gravel and sand particles was observed in test pits dug in the upstream protection zone and that the fine-grained portion of the mixture actually governed the stability of the slope. Based on these data, it appears that the fine-grained portion of the protection zone probably was responsible for the high excess pore pressures and/or liquefaction which lead to slope failure and lateral spreading of the upstream slope of the dam.

Cases 6 and 7 - Pence Ranch and Whiskey Springs, 1983 Borah Peak, Idaho, Earthquake

Liquefaction of gravelly soils in level ground at Pence Ranch (Case 6) and in sloping ground at Whiskey Springs (Case 7) during the 1983 M7.3 Borah Peak, Idaho, earthquake represent a couple of the more well-documented case histories studied to date. Extensive field exploration, including Cone Penetration Tests (CPT), SPT, BHPT, and measurements of in situ shear wave velocities using the Spectral-Analysis-of-Surface-Waves (SASW) Method, as well as sampling of the materials, and liquefaction analyses have been carried out by various investigators at both sites to enable assessment of all the factors pertinent to each case.

In addition to the Pence Ranch and Whiskey Springs liquefaction observations, absence of liquefaction evidence for gravelly deposits is significant as well. In that regard, it is important to note that field reconnaissance conducted by Youd et al. (1985) along a 8 km stretch of the Big Lost River, which lies between 3 and 6 km from the southern terminus of the fault rupture, did not produce any evidence of significant liquefaction effects. They state that "...*the probable reason for the absence of effects in this area is that the sediments may be too coarse and well drained to liquefy or to create surficial liquefaction effects.*"

The Pence Ranch site (Case 6) is located on a low terrace along the Big Lost River approximately 8 km south of the southern edge of the fault rupture. Several gravelly sand boil deposits were observed in the zone of fissuring and lateral spreading after the earthquake. Samples of these deposits were found to be generally clean sand with gravel contents ranging from 5 percent to 25 percent and pebbles as big as 25 mm across. Grain-size distribution curves for samples taken from various sand boil deposits are shown in Figure 3. The subsurface soils at the Pence Ranch site range from a clean gravelly sand (SP-GP) to a clean sandy gravel (GP) with fines content generally less than 5 percent (Site 1 of Figure 3). A zone identified by site investigators as Unit C, located at a depth of about 1.5 to 4 m, had measured blow counts (SPT "N" values) that ranged from 1 to 16, with corrected values, $(N_1)_{60}$, of less than 8. Below this depth, the blow counts were considerably higher (greater than 20). Values of equivalent $(N_1)_{60}$ values were also estimated by Harder (1988) from blow counts obtained using the BHPT. He also concluded that within Unit C, the equivalent $(N_1)_{60}$ is approximately 8; the blow count then increases to 18 at depths between 4 and 6 m.

Evaluations of the liquefaction potential of Unit C were carried out by Andrus and Youd (1989) and Stokoe et al. (1989) using the corrected $(N_1)_{60}$ data, and by Harder (1988) using the equivalent $(N_1)_{60}$ estimated from the BHPT blow counts. Andrus and Youd

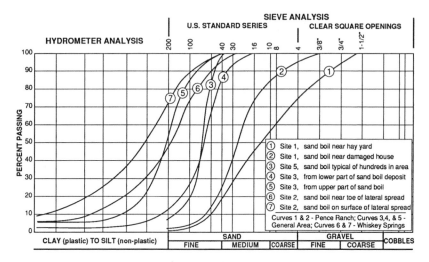

Figure 3. Gradation of Soils Which Liquefied During 1983
Borah Peak Earthquake (Youd et al., 1985)

(1989) estimated that a peak horizontal ground acceleration of 0.30g to 0.35g was experienced at the Pence Ranch site, whereas Harder (1988) had estimated a value of 0.29g. Both studies were able to predict liquefaction of the gravelly sands, as was observed during the 1983 earthquake (Figure 4). It should be noted that the liquefaction analyses conducted by Andrus and Youd (1989) and by Stokoe et al. (1989) were carried out using the procedures developed by Seed et al. (1985) for clean sands without any modifications whatsoever for the gravelly nature of the deposits.

At the Whiskey Springs site (Case 7), field investigations and studies similar to those carried out at the Pence Ranch were also conducted. The Whiskey Springs site is a lateral spread that occurred approximately 2 km from the earthquake epicenter. Estimates of peak ground accelerations at this site range from 0.40g (Harder, 1988) to 0.70g (Andrus and Youd 1989). Because of the rather large lateral movements (roughly 1.2 m) that occurred on a moderately flat slope (approximately 5 degrees), it was concluded that liquefaction was associated with the observed behavior. This was supported by the fact that several sandy silt boils were found near the fissures and sod buckles.

A dense to very dense silty sandy gravel caps the Whiskey Springs site (Units A and B). Looser- and finer-grained gravelly sediments (GM), Units C1 and C3, lie beneath the dense gravel. SPT blow count measurements in a hollow-stem auger casing gave N-values for Unit C1 ranging from 5 to 14. Blow count measurements were made for each 25mm of penetration; these data were essentially uniform in any given test and appear to indicate that the measured N-values were not significantly affected by the presence of gravels.

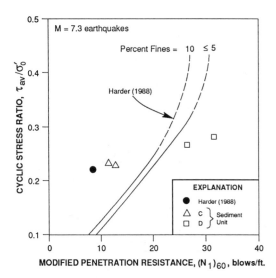

Figure 4. Liquefaction Potential Chart, Pence Ranch Site
(Modified from Andrus and Youd, 1989)

Andrus and Youd (1987) conclude that the lateral spread was caused by liquefaction and shear deformation within Unit C1. The depth of this unit ranges from a minimum of 2.4 m near the toe of the slide zone to a maximum of 16.8 m near the head scarp of the lateral spread zone. Gradation characteristics of soil samples from the suspect liquefied soil layer are presented on Figure 5. An evaluation of the liquefaction potential of the soils underlying the Whiskey Springs lateral spread was performed by Andrus and Youd (1987) using corrected SPT blow counts, simplified procedures developed for sandy soils, and a peak ground acceleration ranging from 0.50g to 0.70g. The results of these analyses are presented on Figure 6.

Harder (1988) performed BHPTs at the same sites investigated by Andrus and Youd (1987). Based on these tests, Harder established equivalent corrected SPT blow counts for the measured Becker Hammer blow counts. The average equivalent $(N_1)_{60}$ for Unit C1 was approximately 8. It was Harder's contention that the results obtained from the SPT investigation gave penetration resistance values that were on the high side. However, he also concluded that the silty gravel layer (Unit C1) was the zone that liquefied and thus triggered the lateral spread. Based on the gradation of these materials (Figure 5), Harder concluded that the larger gravel particles are simply floating in a silty sand matrix having a sand and fines content of 40 percent or more. Harder performed analyses similar but somewhat more elaborate than Andrus and Youd (1987) using an equivalent corrected SPT blow count of 8 and a peak ground acceleration of 0.40 g. Harder's results are also shown on Figure 6; as did Andrus and Youd (1987) these results predict liquefaction of the suspect silty gravel layer (Unit C1).

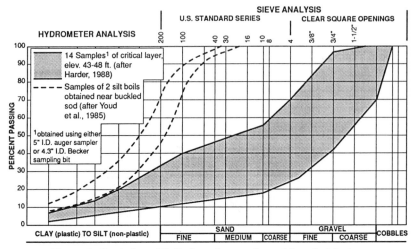

Figure 5. Gradation of Soils from Suspect Layer Liquefied during
1983 Borah Peak Earthquake, Whiskey Springs Site

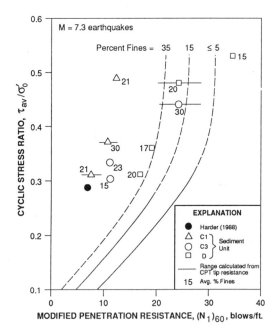

Figure 6. Liquefaction Potential Chart, Whiskey Springs Site
(Modified from Andrus and Youd, 1987)

CONCLUSIONS FROM CASE HISTORIES

Based upon our review of case histories available in the literature, we draw the following conclusions regarding the occurrence of liquefaction of gravelly soils and practical assessments of liquefaction-related issues for sites underlain by gravelly deposits:

1. There is ample evidence that liquefaction of very loose to medium dense gravelly deposits can and does occur; however, SPT blow counts for such occurrences are typically less than 20. For the best documented cases where gravelly deposits have reportedly occurred, SPT blow counts were less than 15.

2. Characterization of penetration resistance in gravelly deposits can be based on a variety of field exploration techniques. For the best documented case histories where gravelly deposits have reportedly occurred, results of liquefaction evaluations that were based on conventional SPT measurements compared favorably with those based on the BHPT. This is true for gravelly deposits containing significant amounts of gravel-sized particles. Therefore, it appears that simplified empirical liquefaction procedures (Seed et al., 1985) developed from sand/silt behavior are generally appropriate for assessing liquefaction potential of gravelly soils.

3. Commonly, the liquefaction behavior of a gravelly deposit is dominated by the finer-grained (e.g., sand, silt) matrix, rather than the gravel particles. In the silty/sandy gravel deposits that liquefied in the past, the gravel-sized particles may simply have been floating in a sand/silt matrix without direct contact with each other.

4. Because of the drainage/dissipation characteristics of gravelly deposits, strong shaking duration must be long enough to sustain accumulation of excess pore pressure; i.e., large-magnitude (M > 7 1/4) earthquakes are required.

5. Given an occurrence or expectation of liquefaction of a gravelly deposit, related ground failure phenomena (e.g., lateral spreading, settlement) are generally less severe than similar liquefaction of sand deposits.

SITE-SPECIFIC ANALYSIS

Liquefaction potential for the saturated sandy and gravelly soils underlying the previously mentioned Central California water treatment facility site (WTF) was evaluated for different postulated earthquake ground motions. The WTF is located very near the San Andreas and Calaveras faults. The site was strongly shaken by the 1989 Loma Prieta earthquake (M_w = 7.0); the peak horizontal ground acceleration recorded during the earthquake at a strong-motion station near the WTF was 0.55 g. Evidence of liquefaction at the site during or after the earthquake was not observed.

The WTF lies in the flat terrain of a flood plain. The site is located at the southern boundary of an area mapped as marshy freshwater basin deposits. The basin deposits are surrounded by alluvial fan deposits consisting of sand, silt and clay.

Over a number of years, the subsurface soil conditions at the WTF have been explored using rotary wash (mud rotary) borings, hollow stem auger borings, and test trench excavations. The subsurface information indicated the WTF is underlain by a thick sequence of alluvial materials that include clay, silt, sand and gravel. Three principal stratigraphic horizons were identified: 1) a 2.7- to 4.3-m-thick surficial clay unit; 2) a 7.6- to 12.2-m-thick unit comprised of predominantly sand and gravel; and 3) a 15.2- to 18.3-m-thick lower unit consisting predominantly of silty and sandy clay.

Groundwater in the confined sand and gravel deposits fluctuate significantly over the year. Water levels measured in the winter rainy season typically are 3 to almost 6 m higher than those measured in the drier summer months.

Liquefaction of the 7.6- to 12.2-m-thick sand and gravel unit was a concern during initial WTF development evaluations. The range of gradations of the gravelly soils underlying the WTF site is presented on Figure 7. The gravelly soil deposits contain between about 20 to 60 percent gravel-sized particles. For analysis purposes, the sand and gravel deposits were assumed to be completely saturated, as is the case throughout the winter rainy season.

Field SPT measurements, visual examination of samples recovered during the field investigations, results of laboratory testing, and review and interpretation of boring logs comprised the various elements of the liquefaction evaluations performed for the WTF.

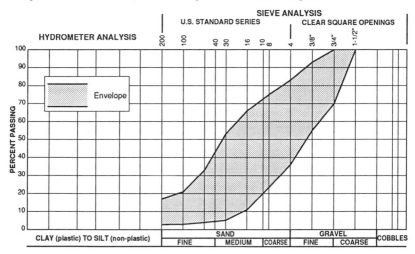

Figure 7. Range of Grain Size Gradations for Gravelly Soils at WTF

Methodology

The methodology used for liquefaction assessment, based on SPT data, is the empirically-based method presented by Seed et al. (1985). Based on the review of the case histories previously described and because measured SPT below counts measured in the gravelly soil deposits were greater than 15 and typical more than 20, it was concluded that the Seed et al. (1985) method would reasonable predict the liquefaction potential of the site soil.

The usual approach when performing a liquefaction analysis is to compute the cyclic stress ratio induced by the design earthquake within the soil deposit, and to compare this value at each depth of interest with that required to cause liquefaction. However, in many cases the end result of a liquefaction evaluation necessitates that remedial measures be considered to improve the in situ density of the soil deposit to a level which would preclude liquefaction. Thus, it is usually of interest to establish the increase in the in situ blow counts required to accomplish this. With this in mind, the liquefaction analyses conducted for the WTF site computed the earthquake-induced cyclic stress ratio developed within the soil deposit, and then used this value and the corresponding empirical liquefaction correlation for clean sand, to establish the critical $(N_1)_{60}$ value, such that soils having a value greater than the critical value would most likely not liquefy, whereas those having a value less than the critical value would more than likely liquefy.

Values of the earthquake-induced cyclic stress ratio were computed using the simplified procedure developed by Seed and Idriss (1970, 1982). Analyses were conducted for a postulated design earthquake on the nearby San Andreas fault and for the motions recorded during the 1989 Loma Prieta earthquake.

Average soil profile conditions and corresponding properties were established based on a review of all available data. Because the site was to be raised by approximately 1.5 m during development, this future ground level was used in the analyses. The ground water level was assumed at a depth of 3 m in all analyses with the exception of the analysis for the Loma Prieta earthquake for which water was assumed to be at a depth of 4.9 m.

Ground water conditions at the time of the field investigations were used to correct all measured SPT blow count data. Total and effective overburden pressures for the assumed conditions were then calculated for the soil profile. Using the computed values of earthquake-induced cyclic stress ratios, corresponding values of the critical $(N_1)_{60}$ were determined using the clean sand (5 percent fines) boundary curve in Figure 8 after adjusting it for the particular earthquake magnitude. The critical blow counts were then compared with the fines corrected $(N_1)_{60}$ blow counts established from the SPT data base to evaluate the liquefaction potential at each location of interest as a function of depth.

Results of Analyses

Comparisons of critical blow counts and clean sand corrected blow counts are presented on Figure 9 for the 1989 Loma Prieta earthquake and for the postulated design earthquake on the San Andreas fault. The open symbols on these plots represent the critical blow counts required to cause liquefaction for each of the earthquakes; the solid symbols are the corrected clean sand blow counts corresponding to the field measured SPTs.

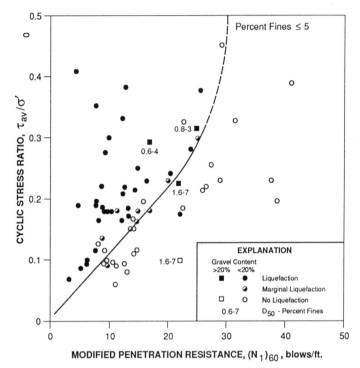

Liquefaction susceptibility chart with data prepared by Seed et al. (1985) for clean sands and a 7.5 magnitude earthquake. Data points having more than 20% gravel are denoted by squares. The mean grain-size and fines content for each of these gravelly sites are also shown.

Figure 8. Liquefaction Analysis

The results of the liquefaction analyses conducted for both earthquakes indicate that some limited liquefaction may have been expected to develop at depths ranging between 6 and 12 m. After the 1989 Loma Prieta earthquake, however, no surface evidence of liquefaction at, or in the immediate vicinity of, the WTF was observed. This may have been due to several factors including the following:

• The site is overlain by a 3-m-thick deposit of clayey materials which could have masked liquefaction effects. Case history studies by Ishihara (NRC, 1985; Ishihara, 1985) have shown that the presence of a non-liquefiable surface layer can prevent the observable effects of at-depth liquefaction from reaching the ground surface.

• Excess pore pressures that may have developed in the subsurface soils during the earthquake may have quickly dissipated due to the generally gravelly nature of the deposit.

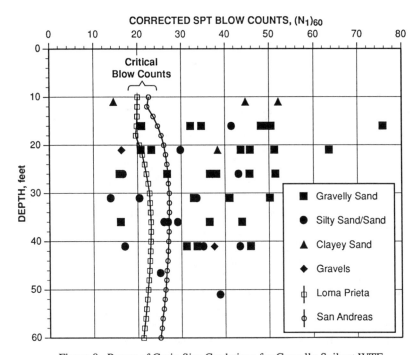

Figure 9. Range of Grain Size Gradations for Gravelly Soils at WTF

- The estimated depth of the water table at the time of the earthquake was about 4.9 m and stress conditions conducive to liquefaction may not have been present.

- The duration of strong shaking of the earthquake was relatively short and high pore pressures were not able to accumulate.

It should be noted that this site has also been subjected to numerous past earthquakes including the M_w8 1906 San Francisco earthquake. No evidence of liquefaction or related effects in the immediate site vicinity was reported for these events (Youd and Hoose, 1978).

Based on the results of liquefaction analyses and the findings of the case history review, it was concluded that the liquefaction potential of the gravelly soil deposits at the WTP are low to moderate for the postulated design earthquake. It was concluded, however, that the consequences of future liquefaction will be relatively minor, should it occur. This conclusion was based on the following considerations:

1. The "corrected" SPT blow counts in the gravelly soil deposits were all greater than 15 and typically greater than 20.

2. The 3-m-thick clayey soil deposits that mantle the site likely will mask liquefaction effects as in past earthquakes.

3. The topography of the site is flat, thus limiting the potential for significant lateral spreading.

It also was concluded that the only consequence of liquefaction would be limited to vertical settlement from dissipation of excess pore pressure in those zones where liquefaction or high excess pore pressures might develop during or after strong earthquake shaking. Estimates of the magnitude of such settlement were on the order of 25 to 75 mm. The WTF was designed to accommodate these settlements.

REFERENCES

Andrus, R. D., and Youd, T. L., 1987, Subsurface investigation of a liquefaction-induced lateral spread, Thousand Springs Valley, Idaho: U.S. Army Corps of Engineers, Geotechnical Laboratory Miscellaneous Paper GL-87-8, 131 p.

Andrus, R.D., and Youd, T.L., 1989, Penetration tests in liquefiable gravels: Twelfth International Conference on Soil Mechanics and Foundation Engineering, Rio de Janeiro, August.

Bartlett, S.F., and Youd, T.L., 1992, Case histories of lateral spreads caused by the 1964 Alaska earthquake: in Case Studies of Liquefaction and Lifeline Performance During Past Earthquakes, Volume 2, United States Case Studies; edited by T.D. O'Rourke and M. Hamada, Technical Report NCEER-92-0002.

Coulter, H.W., and Migliaccio, R.R., 1966, The Alaska earthquake, March 27, 1964: effects on communities, effects of the earthquake of March 27, 1964 at Valdez, Alaska: U.S. Geological Survey Professional Paper 542-C.

Hamada, M., Yasuda, S., and Walkamatsu, K., 1992, Large ground deformations and their effects on lifelines: 1948 Fukui earthquake: in Case Histories of Liquefaction and Lifeline Performance During Past Earthquakes, Volume 1, Japanese Case Studies; Edited by M. Hamada and T.D. O'Rourke, Technical Report NCEER-92-0001.

Harder, L.F. Jr., 1988, Use of penetration tests to determine the liquefaction potential of soils during earthquake shaking: Ph.D. Dissertation, University of California, Berkeley.

Hynes-Griffin, M.E., 1988, Pore pressure generation characteristics of gravel under undrained cyclic loading: Ph.D. Dissertation, University of California, Berkeley.

Ishihara, K., 1985, Stability of natural deposits during earthquakes: Proceedings of the Eleventh International Conference on Soil Mechanics and Foundation Engineering, San Francisco, California, v. 1, pp. 321-376.

National Research Council, 1985, Liquefaction of soils during earthquakes.

Ross, G. A., Seed, H.B., and Migliaccio, R.R., 1969, Bridge foundation behavior in Alaska earthquake: Journal of the Soil Mechanics and Foundation Division, ASCE, v. 95, no. SM4, p. 1007-1036, July.

Seed, H.B., and Idriss, I.M., 1970, A simplified procedure for evaluating soil liquefaction potential: report no. EERC 70-9, University of California, Berkeley, California.

Seed, H.B., and Idriss, I.M., 1982, Ground motions and soil liquefaction during earthquakes: Monograph series, Earthquake Engineering Research Institute, Berkeley, California.

Seed, H.B., Tokimatsu, K., Harder, L.F., and Chung, R.M., 1985, Influence of SPT procedures in soil liquefaction resistance evaluations: Journal of the Geotechnical Engineering Division, ASCE, v. 111, no. 12, p. 1425-1445, December.

Wang, W., 1984, Earthquake damages to earth dams and levees in relation to soil liquefaction: Proceedings of the International Conference on Case Histories in Geotechnical Engineering.

Youd, T.L., Harp, E.L., Keefer, D.K., and Wilson, R.C., 1985, The Borah Peak, Idaho earthquake of October 28, 1983--liquefaction: Earthquake Spectra, Earthquake Engineering Research Institute, v. 2, no. 1, November.

Youd, T.L., and Hoose, S.N., 1978, Historic ground failures in Northern California triggered by earthquakes: U.S. Geological Survey, Professional Paper 993, 177 p.

259

Author Index
Page number refers to the first page of paper